FUNCTIONAL DESIGN

Biological Sciences

—————

Editor

PROFESSOR A. J. CAIN
MA, D.PHIL

Professor of Zoology
in the University of Liverpool

FUNCTIONAL DESIGN
IN FISHES

R. McN. Alexander

Professor of Zoology in
the University of Leeds

HUTCHINSON UNIVERSITY LIBRARY
LONDON

HUTCHINSON & CO (Publishers) LTD
3 Fitzroy Square, London W1

London Melbourne Sydney Auckland
Wellington Johannesburg Cape Town
and agencies throughout the world

First published 1967
Second edition 1970
Third edition 1974

Printed in Great Britain by The Anchor Press Ltd,
and bound by Wm. Brendon & Son Ltd,
both of Tiptree, Essex
ISBN 0 09 104750 1 (cased)
ISBN 0 09 104751 X (paper)

CONTENTS

FIGURES

PREFACE

This book is about the aspects of fish design which I happen to find particularly interesting. Inevitably, some readers will find their interests neglected. I hope that they will forgive me.

I have been helped in writing this book by conversations with my colleagues in the Department of Zoology of the University College of North Wales, and with Dr P. H. Greenwood. Dr P. O'Donald criticised a draft of Chapter 1. Fig. 12 A–C has been traced, with permission, from photographs published in the *Journal of Zoology*.

I

DESIGN BY SELECTION

This book is about relationships between the anatomy, physiology and natural history of fishes. These relationships usually have an appropriateness about them which we attribute to evolution, but which resembles the appropriateness of structure to function characteristic of well-designed implements. The effectiveness of evolution is so well established that where the design of a fish seems bad it is probably only because we have failed to understand it. It is assumed that all the anatomical and physiological characteristics discussed in this book are inherited, and that they have evolved by natural selection.

This chapter is about the kinds of character which will be favoured by natural selection and so can be expected to evolve. One of its objects is to show that natural selection offers a plausible explanation of some of the apparently trivial aspects of fish design which are described in subsequent chapters. For this purpose we need a measure of the extent to which natural selection will favour one genotype rather than another. The customary measure is the coefficient of selection.

Consider a population which includes individuals of genotypes A and B. Zygotes of either genotype may die without becoming adult, or may become adult and yet fail to breed, or may produce gametes which become incorporated in zygotes of the next generation. Suppose that zygotes of genotype A contribute an average of a gametes each to zygotes of the next generation, and that zygotes of genotype B contribute an average of b gametes each to zygotes of the next generation. Then the coefficient of selection s, of A relative to B is given by the equation

$$1 - s = a/b \qquad (1)$$

The selective coefficient is positive. when A is selected against and $a < b$. It is negative when A is favoured by selection.

A gene which is favoured by selection will tend to become increasingly common in a population. The rate of increase will depend on the coefficients of selection of both the homozygote and the heterozygote. We will consider only the particular case where there is no dominance and the coefficient of selection of the heterozygote is half the coefficient of the homozygote. Then if the homozygote G_1G_1 has a coefficient of selection s relative to the homozygote G_2G_2, the heterozygote G_1G_2 will have a coefficient $s/2$ relative to G_2G_2. It can be shown mathematically that in such a case, provided that s is small, the ratio of the number of genes G_1 to the number of genes G_2 in the population will decrease, being divided by e (that is, about 2·7) every $2/s$ generations.* If s is negative, the ratio will of course increase. For instance, if $s = -0·01$, the ratio will be multiplied by about 2·7 every 200 generations.

A genotype will have a negative coefficient of selection if its possessors have reduced mortality or increased fertility. A number of the characters described in this book can be expected to affect the mortality of fish. A character which increases the maximum swimming speed of a fish may reduce mortality by enabling it to escape from predators more often. One which makes it inconspicuous may reduce mortality by decreasing the frequency with which predators notice it. I have no information on fish from which the coefficients of selection of the corresponding genotypes can be calculated, but there is some information about genotypes which affect predation in other animals. For instance, the non-melanic form of the moth *Biston* is more conspicuous than the melanic form when they rest on tree-trunks in industrial areas and is more liable to be eaten by birds. It has been estimated that in such areas its coefficient of selection, relative to the melanic form, is around 0·6.[33]

A large proportion of the characters described in this book

* Dr P. O'Donald has shown me this.
[33] Superior figures refer to References, pp. 150–4.

have no apparent bearing on mortality and no direct bearing on fertility. They affect the quantity of food the fish can obtain or the amount of energy it uses in metabolism. The purpose of the following pages is to show that a genotype which increases the amount of food a fish obtains, or decreases its energy consumption, will tend to increase the number of eggs laid by females and to be favoured by natural selection.

Consider a population of a species of fish, of which some individuals have genotype A and some have genotype B. We will suppose that A females lay on average more eggs each than B females, but that there is no other phenotypic difference between individuals of the two genotypes. Genotype A will be favoured by selection in some circumstances, but not in others.

It might not be favoured if each breeding pair established a territory in which only its offspring fed. If the territories were equal in size offspring of A mothers would be more crowded than offspring of B mothers, and would tend to get less food each. Fewer might survive the period of life in the territory. Genotype A might be selected against although, and indeed because, it increased the number of eggs. In more general terms, genotype A might be selected against whenever the offspring of a given mother competed more with each other than with the offspring of other mothers. Trout (*Salmo trutta*), for instance, bury their eggs in a group in gravel in a stream. The offspring of a mother tend to stay near the place where the eggs were laid, and so to compete more with each other than with other young trout. Some Cichlidae look after their young for some days after hatching,[5] and the offspring of a mother probably compete more with each other, at this stage, than with the offspring of other mothers. In these and similar cases, an increase in the number of eggs laid might be selected against.

When the offspring of different mothers are mixed indiscriminately from the earliest age at which they compete, genotype A will be favoured by selection. If food is limited, fewer young may survive to a given age than if all the mothers had genotype B and laid the smaller number of eggs, but the survivors will include the same proportions of offspring of A and B mothers as the eggs. If A mothers lay 1% more eggs than B mothers, while A and B fathers are equally successful in fertilising eggs, genotype A will have a coefficient of selection, relative to B, of -0.005.

Random mixing of offspring seems to occur in a great many

fish. Nearly all marine fish, apart from those that live on shores, lay eggs which float in the plankton and must get thoroughly mixed.[5] Shore fish such as the blennies (Blenniidae) lay their eggs in clumps, stuck to rocks or seaweeds, but the offspring of different mothers probably get thoroughly mixed shortly after hatching by the action of waves and tides. The argument that follows assumes random mixing.

The number of eggs which is laid can be increased by increasing the size of the ovary relative to the size of the fish, but there must be selection against excessively large ovaries. For instance, a fish with a very large proportion of ovary in its body would have a relatively small proportion of muscle, and would be rather a slow swimmer. It might be at a disadvantage in catching prey or escaping predators. In mature female fish the eggs may be as much as 20% of the weight of the fish. The percentage varies between species but seems to be fairly constant for fish of a given species, provided that the fish are reasonably well fed.[66] When food is short the size of the ovaries may be reduced.

If the size of the ovary is fixed the number of eggs can still be increased by reducing their size, but there must be selection against excessively small eggs. If the egg is small the newly hatched fish must also be small and may be at a disadvantage in competition with larger fry hatched from larger eggs. They may be more vulnerable to predators, or they may suffer from the size hierarchy effect. When fish of the same species but different sizes are kept together in aquaria, the larger ones (i.e. the ones at the top of the size hierarchy) tend to grow relatively faster than the smaller ones. A fish will grow faster in the company of fish smaller than itself than in the company of fish larger than itself, even if plenty of food is available.[1] We will see presently that selection tends to eliminate characters which reduce the growth rate.

There are, then, factors which tend to limit the number of eggs produced by a female of given size. Excessively large ovaries and excessively small eggs are disadvantageous. Within a teleost species the number of eggs produced tends to be proportional to the weight of the body.[1a] Big females produce more eggs rather than bigger ones. A genotype which increases the size of the females at a given age without affecting mortality will be favoured by selection. If females of genotype A grow 1% faster than females of genotype B, they will lay 1% more eggs in the course of their lives and the coefficient of selection of A relative to B will be −0·005.

The average growth rate of fish in a population depends on the amount of food they eat and the amount of energy they use in metabolism. This relationship is expressed by the equation

$$uF = g(G + H) + R + S \qquad (2)$$

The capital letters represent quantities of energy and the small letters represent ratios. F is the energy content of the food eaten annually by the population, and u is the proportion of this energy that is actually utilised, as opposed to being lost in the faeces or urine. G is the amount of energy which is incorporated annually in the bodies of the population by growth and H the energy incorporated in eggs and sperm. These include energy incorporated in fish which die in the course of the year, as well as in those which survive. The incorporation of a quantity of energy $(G + H)$ will require the use of a greater quantity of food energy, since energy is used in feeding and digestion, and since energy is lost from the food in digestion and re-synthesis. This is indicated by the factor g. R and S represent the amount of energy used annually in metabolism for purposes other than growth. S is the amount used in swimming and R is the amount used for other purposes.

If we are to use this equation to predict coefficients of selection, we must be able to assign numerical values to all the quantities in it. We will try to estimate typical values for teleosts.

Teleosts usually lose around 20% of the energy content of their food in their urine and faeces,[71] so we can take $u = 0.8$. The other ratio, g, can be estimated from the results of investigations of the growth of fish. There is a certain daily ration which a fish must eat if it is neither to grow nor to lose weight. This is known as the maintenance ration and can be determined by experiment. If the fish eats more than the maintenance ration it grows. Pike (*Esox*) have been found to increase in weight by about 0·45 g for every 1 g of food they eat in excess of the maintenance ration.[60] They were fed on small fish, which probably had about the same energy content, weight for weight, as the material which they added to their bodies as they grew. Thus, if u was 0·8, g must have been about 0·8/0·45 or 1·8. Experiments with other species have given similar values.[1a] We will probably not be very far wrong if we take $g = 2$.

The resting metabolic rate is the rate at which a resting animal uses oxygen. It depends on many things. A fish which has fed recently uses more oxygen than a starving one, and an excited fish uses more oxygen than a calm one. Resting metabolic rates determined by different investigators for fish of the same species and size at the same temperature may differ by as much as a factor of 2. Estimates of R can be made from determinations of resting metabolic rate, but they are necessarily rough estimates.

Mann determined the resting metabolic rates of several teleosts which are common in the River Thames.[71] He used specimens of various sizes and made measurements at several temperatures. A census had provided estimates of the numbers and sizes of the fish in the river. Mann was able to calculate the amount of energy used annually in resting metabolism by the whole population of each species in a unit area of the river. He took account of the different metabolic rates of fish of different sizes and of the annual fluctuations of the temperature of the river. He was also able to calculate, from the results of the census, the amount of energy incorporated annually in the growth of the population and in eggs and sperm. His calculations thus give values for R, G and H. For these particular species in this particular river, G seems to be 11–18% of R, and H 1–3% of R. We will suppose that these are reasonably typical values, and use them in our calculations.

S, the energy used in swimming, can be estimated from the maintenance ration. For fish on maintenance rations

$$uF \simeq R + S$$

since $G = 0$ in these circumstances and H is always small. Such estimates are likely to be too low. Fish on maintenance rations may adapt to their low diet by reducing the resting metabolic rate or their activity or both below their normal values.[1] Fish kept in aquaria for experiments may be less active than they would be in natural conditions. A calculation based on the maintenance ration of trout (an active fish), gives $S \simeq R$.[71] An ingenious investigation of carp (*Cyprinus*) in a lake contaminated with radioactive wastes gave a similar result. For lack of any sounder basis for an estimate we will assume $S = R$.

We have estimated u and g and the relative sizes of G, H, R and S. We can now put in equation (2) numerical values which seem reasonably likely to be typical of natural populations of teleosts. These values are shown below, each immediately under the corresponding symbol. F is arbitrarily taken as 100.

$$uF = g(G + H) + R + S$$
$$0{\cdot}8 \times 100 = 2(5 + 1) + 34 + 34$$

We will now consider the effects of small changes in some of the quantities in the equation. Suppose that F is not 100 but 101, but that R and S are unchanged. uF is increased by 0·8 and to balance the equation $(G + H)$ must rise by 7% from 6 to 6·4. Since the size of the gonads depends on the size of the body, and the mean size of the bodies of breeding fish depends on the growth rate, a long-term increase in F will tend to increase G and H by equal proportions. A 1% increase in F could thus be expected to lead to 7% increases in G and H. Similarly, a 1% decrease in R or S without any accompanying change in F should lead to 3% increases in G and H.

We can now estimate the coefficients of selection of genotypes whose possessors can obtain 1% more food than fish of another genotype, or use 1% less energy in resting metabolism or swimming. We will, however, have to make two more simplifying assumptions. Both the assumptions are probably false and they are made solely because we do not know enough to make more realistic assumptions; each will tend to make us underestimate the extent to which the genotype is favoured by selection.

First, we will assume that there is no size hierarchy effect. Fish which grew·faster for the reasons we are discussing would be larger, at any given age, than competitors with the other genotype. If there was a size hierarchy effect their greater size would itself give them an advantage in obtaining food and they would grow faster still. Secondly, we will assume that the size of males does not affect their success in fertilising eggs. In fact a large male might have a better chance of finding a mate than a smaller one and, since it produced more sperm, might be able to fertilise more eggs.

If fish of genotype A can obtain 1% more food than fish of genotype B, they can be expected to lay on average 7% more eggs each and the coefficient of selection of A relative to B can be expected to be $-0{\cdot}035$. If fish of genotype A′ use 1% less energy in resting metabolism or in swimming than fish of genotype B, they can be expected to lay on average 3% more eggs each and the coefficient of selection of A′ relative to B can be expected to be −0·015. These are very rough estimates based, as we have seen, on numerous assumptions.

An example will show the use that can be made of these estimates. Another rough calculation in Chapter 4 leads to the conclusion that the energy used by a teleost in pumping water over its gills accounts for about 0·5% of its resting metabolic rate. In many species the relative rates of different phases of the pumping movement seem to be adjusted, albeit rather roughly, so as to minimise the energy used. Can these adjustments be explained in terms of natural selection, or are the savings of energy which are involved too trivial?

Suppose a gene appeared in a population which, when homozygous, reduced the energy used in pumping water over the gills by 10% of its amount in resting fish. R would be 0·05% less for the homozygotes than for fish which lacked the gene. The coefficient of selection of the homozygote could be expected to be −(0·05 × 0·015) or −0·0008. We will suppose that there was no dominance, and that the initial frequency of the gene was 1%. It could be expected to rise to 2·7% in about 2/0·0008 or 2,500 generations and to 99% in about 23,000 generations.* 23,000 generations might perhaps take 100,000 years. This is a very short time compared to the time which has been available for evolution. The teleosts appeared over 100 million years ago.[7] The modest effect of the gene would be sufficient to ensure that its frequency increased enormously in a reasonably short time. A 10% reduction of the energy used in pumping was assumed arbitrarily since we do not know how much effect the genes that might be available would have. Nevertheless, the calculation indicates that it is not unreasonable to explain the relative rates of the different phases of the pumping movement as having evolved to minimise the energy used. In more general terms, we may conclude that evolution by natural selection offers a plausible explanation for adaptations which result in small savings of energy.

* When the gene frequency changes from 1% to 99% the gene ratio changes by a factor of about $e^{9·2}$, from 1/99 to 99/1.

2

SWIMMING

The swimming mechanisms of fish can be expected to evolve so as to decrease the energy required for swimming, to increase the maximum acceleration and speed of the fish and to improve manœuvrability. Changes which decrease the energy required for swimming will be favoured because they will make more energy available for growth and so increase the number of eggs which are produced (Chapter 1). Improved speed and manœuvrability will help to enable the fish to escape from predators and they may also enable it to catch more food, if it takes active prey, and so increase its growth rate and the number of eggs it lays.

When a body moves through a fluid a hydrodynamic force acts on it.[14a] A component of the force acts backwards along the direction of motion; it resists the progress of the body and is known as the drag. In some cases the whole of the hydrodynamic force is drag: for example, when a symmetrical body moves along its axis of symmetry or a flat plate moves in its own plane. In other cases the force also has a component at right angles to the direction of motion. This is known as the lift because it is a force of this type, acting upwards on the wings, that keeps an aeroplane in the air. When a flat plate moves along an axis which makes a small angle with its plane the drag is greater than if it moved in its plane, but there is a lift which is several times as big as the drag. The angle between the axis and the plane is known as the angle of attack.

When a fish swims drag acts on its body and must be overcome by a forward propulsive force.

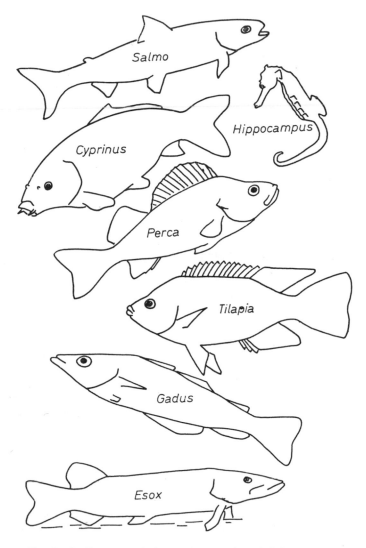

Fig. 1 Outlines traced from photographs of living teleosts: juvenile salmon (*Salmo salar*), carp (*Cyprinus*), sea horse (*Hippocampus*), perch (*Perca*), *Tilapia*, pollack (*Gadus pollachius*) and pike (*Esox*). The photographs were taken in the aquarium of the Zoological Society of London

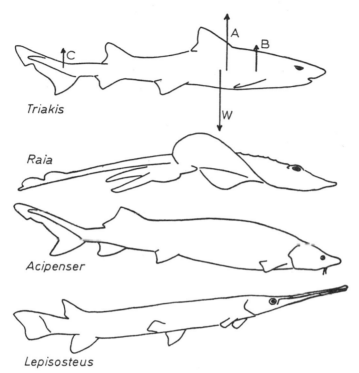

Fig. 2 Outlines traced from photographs of some fish which are not teleosts: leopard shark (*Triakis*), ray (*Raia*), sterlet (*Acipenser ruthenus*) and garpike (*Lepisosteus*). The arrows superimposed on the outline of *Triakis* represent forces, and are discussed on p. 34. The photographs were taken in the aquarium of the Zoological Society of London

Swimming with the tail

Fig. 3A represents a typical fish swimming. As it moves forward through the water it moves its tail from side to side. The tail therefore takes a sinuous path through the water. The forces acting on the tail bend it, but it nevertheless makes an angle with its path through the water.[52] Lift (L_t) therefore acts on it as well as drag (D_t). As the figure indicates, the resultant of the lift and drag on the tail is inclined forwards. When the tail is moving to

21

the left the resultant force on it is inclined to the right and vice versa, but the mean force on the tail, measured over a whole cycle of tail movements, acts directly forward. When the fish is swimming steadily this mean propulsive force must be equal to the drag on the body of the fish (D_b).

Power, or rate of working, is the product of velocity and the force resisting motion. A swimming fish does work against the drag on its body and the drag on its tail. The power required to overcome the drag on the body is the product of this drag and the speed of the body through the water. The power required to overcome the drag on the tail is the mean value, taken over a cycle of tail movements, of the product of the drag on the tail and the speed of the tail through the water.

When the body bends not only does the tail move to the side but the head moves a little to the side as well; the tail wags the fish. This increases the power required for swimming for two reasons. First, work is done against the drag which resists the transverse movements of the head. Secondly, the transverse movements of the head produce a backward force which resists the forward force produced by the transverse movements of the tail.

Natural selection must tend to ensure that the total power required for swimming is as low as possible at any given speed, so that as little energy as possible is needed for swimming and so that the maximum speed which can be attained with muscles of limited power output is as high as possible.

The power used against drag on the tail can be reduced by increasing the ratio of the lift on the tail to the drag on the tail. Less power will then be needed for any given lift. The lift-drag ratio of a fish's tail, or of the wing of an aeroplane, depends on its shape. The ratio of the wing span of an aeroplane to the mean width, from front to back, of its wings is known as the aspect ratio. The lift-drag ratio is highest when the aspect ratio is high.[91] The tails of tunnies and other Scombridae which swim perpetually have high aspect ratios: that is to say, the ratio of the height of the caudal fin to the mean distance between its anterior and posterior edges is high (fig. 3B). The power required to overcome the drag on the tail must be correspondingly low. Most other teleosts have caudal fins of lower aspect ratio, but they use them for hovering in mid-water as well as for forward swimming and their shape may be a compromise between the ideal shapes for

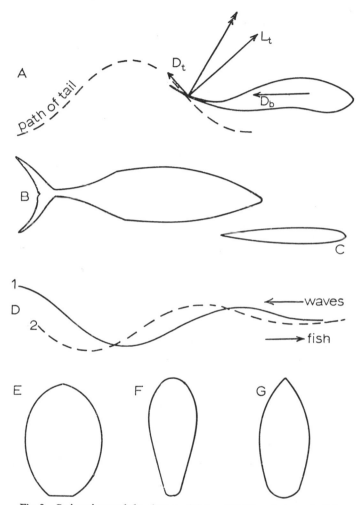

Fig. 3 Swimming and the shapes of fishes. (A) Dorsal view of a fish swimming, showing some of the forces involved. (B) Outline of the body and caudal fin of a tunny (*Thunnus*). (C) Horizontal section through one lobe of the caudal fin of a tunny. (D) Two successive positions of the vertebral column of a fish swimming by means of waves which travel posteriorly along the body, increasing in amplitude as they go. (E), (F), (G) Transverse sections of (E) a mackerel (*Scomber*), (F) a herring (*Clupea*) and (G) a perch (*Perca*)

these two purposes. The manner in which the fins are used in hovering will be described later in this chapter. Some sharks such as *Lamna* swim all the time and have caudal fins very similar in shape to those of Scombridae.

The lift-drag ratio of a fin or aerofoil depends on its shape in section as well as its aspect ratio. The best sections for large, fast-moving fins are streamlined; that is, they are rounded at the front edge, taper to a fine point at the rear, and are absolutely smooth. The caudal fins of Scombridae and sharks correspond quite closely to this ideal (fig. 3C). The best sections for smaller, slower-moving fins with lower Reynolds numbers (see p. 25) are thin arched plates, and smoothness is less important.[14a] Scombridae have more or less rigid caudal fins with tightly packed rays and fairly smooth surfaces. Most other teleosts have caudal fins with uneven surfaces, because their fin rays are separated by very thin webs of skin. We shall see presently that the thin webs are essential to the mobility of the fin and to its use in hovering.

The movements of a swimming fish are not really as simple as we have supposed so far. It does not simply bend its trunk from side to side: the body is thrown into transverse waves which pass posteriorly along the body, increasing in amplitude as they go (fig. 3D). This is most obvious in long slender fish such as the eel (*Anguilla*), where the amplitude of the waves is quite large even at the anterior end of the body. In more typical fish the amplitude of the waves is very small for most of the length of the body but increases sharply where the body tapers near the tail. It is hard to see that such a fish is not simply wagging its tail from side to side, but it is clear from photographs.[41, 42, 52]

Lighthill[67] has discussed the mechanics of this type of swimming. His paper involves mathematical reasoning which many zoologists would find difficult, and only his conclusions will be summarised here.

He showed that the power required for swimming at a given speed would be less when waves of bending travelled posteriorly along the body than when the body simply bent from side to side; it would be less when the amplitude of the waves increased gradually as they travelled backwards, than when their amplitude was constant all along the body. He thus confirmed that fish swim in such a way as to keep the power required low. He went on to show that the power required would be least when the speed at which the fish travelled forward through the water was only a little less than the speed at which the waves travelled posteriorly along the body: that is, when the waves only moved very slowly backwards relative to the water. This could only be achieved when the span of the caudal fin was reasonably big. This is, of course, consistent with our earlier conclusion that a high aspect ratio is advantageous.

Films of various fish swimming show waves passing posteriorly along the body at speeds which are 1·5–2·0 times the forward speed of the body through the water.[41] Lighthill's calculations suggest that the power exerted by these fish would be 1·2–1·3 times the theoretical minimum, which is the product of the swimming speed and the drag which would act on the fish if it was moving at that speed with its body straight. However, this calculation does not take account of the wastage of power due to the tail wagging the fish. A fish which swam by simply bending its trunk from side to side would have to exert at least twice the theoretical minimum power.

We will now consider the drag on the body of the fish which we have just used to define the theoretical minimum power. Part of it is friction drag due to work done against viscosity in the thin layer of water immediately surrounding the fish, while part is pressure drag due to the energy lost as eddies in the wake.

The size of the friction drag depends on the surface area of the body, on its speed and on the nature of the flow of water over it.[14a, 91] The nature of the flow can be predicted from the Reynolds number. The Reynolds number for the flow of water over a body L cm long moving at V cm/sec is $100LV$. When it is less than about 10^6 the water flows smoothly over the body (this is called laminar flow), but when it is more than about 10^6 the flow is turbulent. Flow is probably usually laminar round small fish, but may be turbulent round large fish which are swimming quickly. For a body L cm long with surface area kL^2 cm², travelling in water at V cm/sec, the friction drag is $0·07\,kL^{1·5}\,V^{1·5}$ dynes when flow is laminar, and $0·015\,kL^{1·8}\,V^{1·8}$ dynes when it is turbulent. The quantity k used to specify the area is of course the same for bodies of the same shape but different sizes. The pressure drag cannot be expressed by any simple formula. It depends on the shape of the body and on its position relative to the direction of motion. It can be greatly increased by small protruberances.

For any given body moving in a given direction, it is proportional to the square of the velocity. Thus friction drag is proportional to either $V^{1.5}$ or $V^{1.8}$ and pressure drag to V^2. The power exerted against total drag is the product of the drag and the velocity, and so must be proportional to something between $V^{2.5}$ and V^3. When a fish doubles its speed it can be expected to have to increase its power output 5·6–8·0 times.

Natural selection must tend to favour shapes which give fish the least possible drag at any given speed. Friction drag depends on surface area and so is less for a short plump body than for a long slender one of the same volume. Pressure drag on the other hand is less for a slender body, moving along its long axis, than for a plump one. Experiments with models of airship hulls in wind tunnels have shown that a body of given volume moving at a given speed suffers least drag if it is streamlined and about 4·5 times as long as its maximum diameter. This is the best compromise between the shapes which give least friction drag and least pressure drag. The factor 4·5 is not very critical; the total drag is only 10% higher when the factor is 3 or 7.[14a]

Many fish have bodies whose shapes correspond quite closely to the ideal shape. They include Scombridae (fig. 3B, E) and pelagic sharks. They are smooth and streamlined with no excrescences apart from the fins. They are about 3·5 to 5·5 times as long as their maximum diameter.

Many teleosts have their body flattened from side to side so that its transverse section is a tall ellipse. They are said to be compressed. Their surface areas are greater than if they were round in section, so they presumably suffer more drag. Many pelagic teleosts are silvery and many, like herring (*Clupea*), are also compressed (fig. 3F). Denton and Nicol[35] have shown how silveriness can make a fish inconspicuous when it is swimming in mid-water. A silvery fish is least conspicuous to animals looking at it from greater depths, if it is compressed. Herring are eaten by larger fish and by whales,[3] and may have evolved their shape for this reason. It may be the best possible compromise between the ideals for swimming and for camouflage. Typical acanthopterygians have short compressed bodies which will be discussed at the end of this chapter. Bottom-living fish tend to be flattened dorsoventrally (depressed). One of the consequences of this shape is to make them less conspicuous when they rest on the bottom,

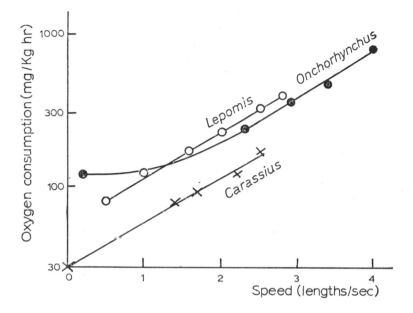

Fig. 4 Graphs of oxygen consumption against swimming speed for *Lepomis*,[28] *Onchorhynchus*[27] and *Carassius*[23, 85]. Further details are given in the table on p. 28

as the areas of their shadows are reduced. Other, more peculiar, shapes must increase drag further. Streamlining must have had very little importance in the evolution of the bizarre shape of the sea horse (*Hippocampus*, fig. 1).

The drag on the bodies of dead fish has been measured in a stream of water in a water tunnel.[77] A live trout (*Salmo*) has been filmed, gliding to a halt with its body straight, and the drag has been calculated from its deceleration.[42] In each case the drag was found to be about seven times the calculated friction drag. The drag on a well-designed airship hull is only about 1·3 times the friction drag.[95] The total drag on the fish seems surprisingly high. The drag on the dead fish may have been unnaturally high because they would tend to flap like flags[52] and the live trout may have been braking slightly.

The metabolic rates of a few teleosts have been measured

Functional Design in Fishes

while they were swimming. In most of the experiments the fish was made to swim against a current of water, and the fall in the oxygen content of the water, due to its respiration, was measured. The metabolic rates found at low speeds varied a lot as the fish were apt to be restless, but graphs can be drawn of the minimum metabolic rate, at a given speed, against the speed. Fig. 4 shows a selection of such graphs. Note that the scale of oxygen consumption is logarithmic. Speed is shown in body-lengths per second. The difference between the metabolic rate at velocity V and the rate at rest is roughly proportional, for the different species, to $V^{1.7}$ to $V^{2.7}$. This is not inconsistent with the conclusion on p. 26 that power output should be proportional to $V^{2.5}$ to V^3, because the efficiency of muscle depends on the rate of contraction.[14a]

TABLE

Some details of the experiments which gave the results displayed in fig. 4, with estimates of the power used in swimming at the stated speeds

	Lepomis	Oncho-rhynchus	Carassius
Weight of fish (g)	30	30	87
Length of fish (cm)	12	16·5	16
Temperature (°C)	20	10	20
Speed to which the following data apply (lengths/sec)	2·8	4	2·5
Oxygen consumption (mg/Kg hr)	400	800	170
Estimated power released by metabolism, x (ergs/sec)	4.9×10^5	9.8×10^5	6.1×10^5
Estimated power used against friction drag, y (ergs/sec)	1.2×10^4	6.2×10^4	2.4×10^4
100y/x	2·5	6	4

This table shows some of the information used to construct fig. 4. It shows the highest speed at which the metabolic rate of each species was measured, and the corresponding metabolic rate. The metabolic rates have been used to calculate the power released by metabolism. (Respiration involving 1 mg oxygen always releases about 3·5 calories or 1.5×10^8 ergs, whatever the food that is being metabolised.) The power exerted against friction drag has been calculated from the size and speed of the fish. The friction drag was calculated from the formula which has already been given for laminar flow, using rough estimates of the area of the body and fins. The power was obtained from it by multiplying it by the speed. It appears from the table that the power required to overcome friction drag is only 2·5 to 6% of the total power released by metabolism. The principal reasons for this seem to be as follows.

(1) A proportion of the power is being used for purposes other than swimming. This proportion is probably small, as the metabolic rate is so very much higher at the speeds in question than in resting fish (fig. 4).

(2) Only part of the chemical energy released by metabolism in the muscles is converted to mechanical work. The efficiency of conversion of chemical energy to mechanical work in human muscles is about 10–20%, according to the speed of contraction. Evidence that the efficiency of fish muscle is similar has been obtained by experiments with fish swimming in a water tunnel.[93a] Objects were attached to a fish to increase the drag and the change in its metabolic rate was noted.

(3) The drag on a swimming fish is probably about three times as great as the drag on a rigid body of the same shape, because the swimming movements reduce the thickness of the boundary layer.

The estimated power required to overcome friction drag is 6% of the power released by metabolism in *Onchorhynchus* (the Pacific salmon) and 4% in *Carassius* (the goldfish). If their muscles have the expected efficiency of 20% or less, the total drag on these species cannot be more than about 3 and 5 times the estimated friction drag. These factors are smaller than the ones calculated from measurements on dead and gliding fish.

The speeds at which fish can swim have been measured in various ways, but the most thorough and accurate measurements have been made by Bainbridge in his 'fish wheel'.[15] This is a large annular trough in which the fish swims. As it swims the trough is rotated in the opposite direction by a powerful motor, so that the fish is kept in the field of a cine camera. The water in the trough is forced to rotate with the trough by an ingenious system of gates which only open as the fish approaches them. A meter which shows the speed of rotation of the trough is photographed at the same time as the fish. The speeds of brief darts as well as of more sustained swimming can be obtained from the film record.

Only a few species of teleosts have been investigated in the fish wheel, but their speeds agree fairly well with speeds measured by other (usually less reliable) methods. Fish between about 4 and 30 cm long have been investigated in the fish wheel, and other methods have given speeds for fish up to 150 cm long. A teleost, whatever its size, can usually attain a speed of about 10 lengths/sec in a short dart lasting a few seconds, but when it swims for several minutes it can only maintain 3–5 lengths/sec.[15] The measurements of metabolic rate (fig. 4) only extend up to this maximum cruising speed. We will see in the next section of this chapter that the higher speeds which cannot be maintained depend on anaerobic metabolism—that is they depend on energy obtained by processes which do not require oxygen. The differences in speed (in lengths/sec) which have been found between different species of teleosts are remarkably small. Selachians probably cannot swim as fast; of the few records available, none is greater than 4 lengths/sec.

Muscles

The swimming muscles of fish are of two types, of which one is typically red and the other white.[25] They have been most thoroughly studied in the dogfish, *Scyliorhinus*, in which, as in most other fish, the red fibres form a thin layer lateral to the much greater bulk of white fibres. Dogfish whose brains have been destroyed can be kept alive by running aerated sea water through their gills. In this condition they make slow spontaneous swimming movements at about the same frequency as the movements of normal dogfish cruising round an aquarium. They can also be stimulated to more vigorous activity. It can be shown, by inserting electrodes into the red or the white muscle and recording its electrical activity, that when the dogfish are making slow swimming movements the red muscles are active and the white ones are not. When they make vigorous swimming movements, the white muscles are active, and it is not certain whether or not the red ones are as well. These experiments and some with funnies suggest that intact fish rely on their red muscles in cruising, and that the white muscles are used only as an additional source of power for bursts of fast swimming.[79b]

Prolonged swimming must be supplied with energy by an aerobic process, since an oxygen debt cannot be allowed to accumulate indefinitely. The red fibres function aerobically, so their power output depends on a sufficiently rapid supply of oxygen. They differ from the white fibres (which function anaerobically) in having a richer blood supply, in being more slender and in containing myoglobin. Slenderness allows faster diffusion of oxygen. Myoglobin, which gives them their red colour, transports oxygen to the oxidising systems. They obtain their energy mainly by oxidising fat which gives twice as much energy, weight for weight, as carbohydrate or protein.

Anaerobic metabolism is preferable for a short burst of speed since its rate is not limited by the rate at which oxygen can be supplied. It makes it possible to attain a higher speed with a given gill area and blood supply. The white muscles obtain their energy by the anaerobic conversion of glycogen to lactate. Two minutes of violent activity may reduce their glycogen content, in a dogfish, by 50%, whereas many hours of slow swimming movements (in which they are not used) leaves their glycogen content unchanged.

We have seen that teleosts can achieve a speed of 10 lengths/sec in a short burst, but can maintain only 3–5 lengths/sec for long periods. The power required at the higher speed must be obtained by anaerobic metabolism in the white fibres. It is not clear why this higher speed cannot be maintained for more than a few seconds. *Onchorhynchus* fatigued by swimming a little faster than the cruising speed can develop an oxygen debt about equal to the maximum amount of oxygen which can be taken up at the gills in 40 mins.[27] It seems most unlikely that so large a debt could be developed in a few seconds even at the highest speeds.

The maximum burst speed is 2–3 times the cruising speed. We have seen that the power required for swimming increases very rapidly with speed. The power output at the burst speed must be much more than 3 times the output at the cruising speed, and a much greater quantity of muscle must be active at the burst speed than is necessary at the cruising speed. There seems always to be much more white muscle than red muscle. The exact proportions vary and can often be related to the habits of the fish. Relatively large proportions of red muscle are found in tunnies, which cruise continually in the sea, and in *Salmo*, which undertakes extensive migrations. Very few red fibres are present in the trunk muscles of pike (*Esox*) and miller's thumb (*Cottus*). *Esox* lies in wait for prey and catches it by a sudden dash. *Cottus* swims slowly by means of its pectoral fins (and has red pectoral fin muscles) and uses the trunk muscles mainly for fast swimming.

Within certain limits biochemical processes occur faster at higher temperatures. A warm muscle can produce more power than a cold one. Most fish have almost exactly the same temperature as the water in which they live, but the red muscles of tunnies (*Thunnus*), and parts of their white muscles, are commonly about 10 centigrade degrees warmer than the water. If the blood from these muscles were allowed to travel still warm to the gills so much heat would be lost there that the muscle temperature could not be maintained. Instead, the arteries and veins of the muscles run parallel to each other, closely packed, so that the blood leaving the muscles warms the blood arriving at them, and very little heat is lost.[31]

The trunk muscles of fish are segmented and each segment has its own nerve supply. Some such arrangement is essential for an animal which swims by passing waves of curvature along its body. The myotomes, or muscle segments, have a very complicated shape which has never been explained satisfactorily.

Selachian fins

Fins are segmented structures. Their radial muscles develop by separating off the myotomes and are served by branches of the appropriate spinal nerves, but the segments are often more crowded in the fin than in the body. A selachian dorsal fin derived from 14 segments may only extend over 6 myotomes.

Fig. 5 shows a section across a typical selachian fin and a side view of a dissection of three of its segments. In each segment there are two cartilaginous radials (some fins have more) joined together by ligaments. The more distal radial lies between two sheets of tightly packed ceratotrichia which extend to the edge of the fin and give it some stiffness. The ceratotrichia consist of a form of collagen known as elastoidin. There is a considerable number of them on each side in each segment, but there is only one muscle on each side in each segment. It inserts by a broad tendon on the ends of the ceratotrichia. When it contracts, it bends

---

Fig. 5   The structure of the dorsal fins of a selachian (above) and a teleost (below). Dissections, seen from the left side, are shown on the left, and transverse sections are shown on the right. *c*, ceratotrichia; *d*, depressor muscle; *e*, erector muscle; *i*, inclinator muscle; *lm*, longitudinal swimming muscle; *r*, radial; *rm*, radial muscle

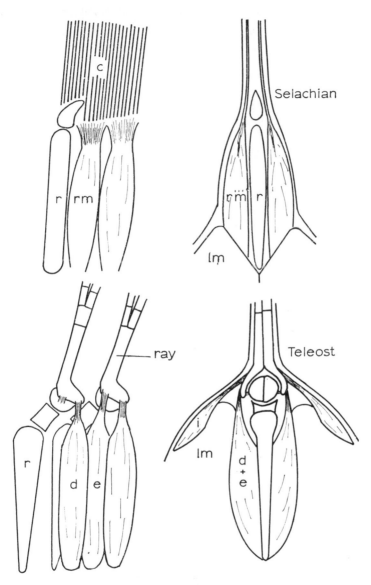

Selachian

ray

Teleost

Fig. 5

the fin to its side. There is no means of altering the area of the fin.

The pectoral fins are more mobile than the other fins of sela-chians. In typical, shark-like selachians (Pleurotremata) the radials are attached to a group of basal cartilages which articulate with the pectoral girdle. The articulation allows the fin to swing forwards and back in its own plane. As the fin swings forward its leading edge turns down and as it swings back its leading edge turns up. Movements at the articulation therefore affect its angle of attack. We shall see presently that they are important in swimming. There are muscles to control them.

Selachians have no swimbladders and are denser than the water in which they live (Chapter 3). They must therefore exert a force against gravity when they swim horizontally. Sharks and dogfish use their pectoral and caudal fins to produce this force.[9] The pectoral fins are held at an angle of attack so that an upward lift acts on them. The caudal fin is of the type which is called heterocercal. Most of it lies ventral to the posterior end of the vertebral column and trunk muscles (see *Triakis* in fig. 2). As the tail moves from side to side through the water this ventral part of the caudal fin bends passively and trails behind. The fin is therefore not vertical as it moves through the water and the lift which acts on it is inclined upwards. The lift has a vertical com-ponent which acts against gravity as well as the horizontal component which provides the propulsive force.

The vertical forces acting on a shark swimming horizontally are shown in fig. 2. The weight of the fish, W, acts through the centre of gravity. The upthrust A, equal to the weight of water the fish displaces, acts through the centre of buoyancy, which lies a little anterior to the centre of gravity as the tail is denser than the head. B is the lift due to the pectoral fins and C the upward component of the lift due to the caudal fin. Since the fish is swimming horizontally the upward forces must equal the down-ward one:

$$A + B + C = W \qquad (3)$$

Also the moments of the forces about the centre of gravity must balance. If the distances from the centre of gravity of the lines of action of A, B and C are a, b and c, respectively,

$$Aa + Bb = Cc \qquad (4)$$

(3) and (4) are a pair of simultaneous equations. A, W, a, b and c are physical properties which are more or less fixed for any given fish (b can be changed a little by swinging the pectoral fins forward and back, and b and c will change slightly as the body bends, but these changes are small and will be ignored). For given values of A, W, a, b and c there is only one value of B and one value of C which will satisfy both equations. Any particular shark requires a fixed upthrust B from its pectoral fins and a fixed upthrust C from its tail, whenever it swims horizontally. B seems to be about 3·2% and C about 1·3% of the body weight for dogfish (*Scyliorhinus*) and tope (*Galeorhinus*). Some selachians are less dense than these ones and would not need such large forces (Chapter 3).

The upward forces produced by the tails of various selachians have been measured in crude experiments in which the tails have been moved transversely at various speeds under water.[9] The force increases as the speed increases and is roughly proportional in every case to the 1·4 power of the transverse speed. The lift produced by an aerofoil or hydrofoil at a fixed angle of attack is proportional to the square of the speed. The angle of attack of the tail is not fixed but decreases as the speed increases since the forces which bend it are greater at higher speeds. It was therefore to be expected that the lift would increase when the speed increased, but that it would not increase as rapidly as the square of the speed.

The tail of a tope gave the required lift of 1·3% of the body weight when its transverse speed was 0·4 times the length of the intact fish per second. The other experiments were done with tails which had been deep-frozen and which were probably more flexible than fresh ones. I have since done some experiments with fresh dogfish and smooth hound (*Mustelus*) tails. Lifts of 1·3% of the body weight were obtained at 0·7 to 1·0 body-lengths/ sec. The mean transverse speed of the tail of a swimming selachian seems to be about equal to the forward speed of the body. If the tail acted passively as in the experiments it would only give the correct lift for horizontal swimming at one critical speed, which would vary from species to species but would be between about 0·4 and 1·0 body-lengths/sec. When a selachian swims horizontally at any other speed it probably adjusts the angle of attack of the caudal fin by means of the fin muscles. Dogfish in aquaria commonly cruise at about 0·5 body-lengths/sec, which seems to be a little below their critical speed.

Harris[48] confirmed that the pectoral fins of *Mustelus* are capable of producing at least the lift required of them, by experiments in a wind tunnel. He used a cast of a specimen which was 35 ins long and probably weighed about 5 lb, and measured the forces on the intact cast and on the cast with various fins removed. From these forces one can calculate the forces which would act on the fins when the fish swam in water. At the angle at which they were fixed to the cast the pectoral fins would have given the required lift of about 3·2% of the body weight at about 0·5 body-lengths/sec. A swimming fish could adjust the angle of attack of the fins to suit its speed. The experiments were done with the pectoral fins in just one position within the range of possibilities.

At this point it is necessary to introduce the concept of hydro-dynamic stability. Consider an arrow tailed with feathers. Suppose that it is somehow deflected slightly in flight so that the length of the arrow is inclined at an angle to its path through the air: the feathers will then produce a lift which will tend to swing the arrow back into the line of flight; deflections tend to be corrected; the arrow is stable. This is only true as long as it is fired in the intended way. If it were fired feathers first any de-flection would be magnified, the arrow would tend to turn about and it would be unstable. An arrow or a fish can be deflected in any direction, but any possible deflection can be resolved into a pitch (up and down) component and a yaw (sideways) com-ponent (fig. 6).

Harris tested the stability of his cast of *Mustelus* in the wind tunnel. The body of the cast was straight and it was naturally not possible to reproduce the sinuous movements of a swimming fish. As he pointed out, one must be a little cautious in applying his findings to living fish. The cast was unstable in pitch. If swimming selachians are similarly unstable the least deflection up or down must tend to be magnified, and then can only maintain a level path by continual up and down steering movements. The pectoral fins are presumably used for this. The cast was neither stable nor unstable in yaw. The anterior dorsal fin tended to make it unstable and the other median fins tended to make it stable, and these two effects were nicely balanced. Harris suggested that the fish might be able to control stability in yaw. A fin will have less effect on stability if it is limp than if it is stiff because it will tend to bend into line with the water flowing over it. The fins of the cast were stiff. Living fish can alter the stiffness of their fins by altering

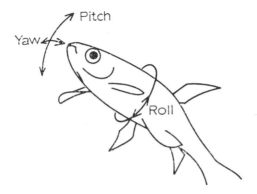

Fig. 6 A teleost seen obliquely from in front and below,
to show the meanings of the terms pitch, yaw and roll

the tonus of the fin muscles. In straight swimming stability is
advantageous because deflections are automatically corrected.
Mustelus might make itself stable in yaw, when swimming
straight, by allowing the anterior dorsal fin to go limp. In turning,
instability is advantageous, because a turn, once started, is automatically assisted by it. *Mustelus* might make itself unstable
in yaw when it turns by stiffening the anterior dorsal fin and
letting the other median fins go limp.

The anterior dorsal fin only tends to make *Mustelus* unstable
because it lies anterior to the centre of gravity. Its position is
unusual. Nearly all other genera of selachians seem to have it
immediately over, or posterior to, the centre of gravity.[70] *Mustelus*
may be able to make itself unstable in yaw, but other selachians
are less likely to be able to do so.

All the fins must tend to discourage rolling (rotation about the
direction of motion). When the fish starts to roll, the path of each
fin becomes helical instead of straight. The fin thus comes to have
an angle of attack and produces a lift which resists the roll.

All the fins serve their functions by acting as hydrofoils. Their
lift is useful but their drag hinders swimming. We have seen that
the drag of a hydrofoil can be reduced by giving it a streamlined
section and a high aspect ratio. Selachian fins are streamlined in
section, and the pectoral and caudal fins of pelagic sharks such
as *Lamna* have high aspect ratios.

37

Rhinobatos is a dorso-ventrally flattened selachian with very large pectoral fins (fig. 7A). They extend anteriorly.dorsal to the gill openings and posteriorly almost to the pelvic fins. When this fish is travelling slowly it swims with its tail in the usual way but when it swims fast it moves its pectoral fins as well.[64] They are thrown into waves which travel posteriorly along the fin increasing in amplitude as they go. These waves have a propulsive effect just like that of the usual transverse waves of the body though they involve vertical instead of transverse movements.

In rays such as *Raia* the pectoral fins are even bigger; the part of the body posterior to the anus is slender and the dorsal, anal and caudal fins are small or lost. The tail is not used in swimming which depends entirely on undulation of the enormous pectoral fins (fig. 2).

Rays, like sharks, are denser than sea water and need an upward lift when they swim. Their shape, with a flat ventral surface and a convex dorsal one, should generate an upward lift even at zero angle of attack, and this lift may be sufficient but it does not seem to have been measured.

Fins of bony fish

In the fins of bony fish the place of the ceratotrichia is taken by jointed bony rays, each of which consists of a half ray on each side of the fin (fig. 5). The rays branch distally. The primitive palaeoniscoids had the rays closely packed in their fins, and the sturgeon (*Acipenser*) resembles them. It has two or three rays in each segment of each fin, fin muscles arranged like those of sharks and the same restricted range of fin movements as sharks. It has a small swimbladder but is nevertheless denser than sea water, and has a heterocercal tail (fig. 2.) The palaeoniscoids and other early bony fish had heterocercal tails.

The fins of holosteans and teleosts are very different from those of palaeoniscoids and sturgeons. There is only one ray in each segment of each fin (fig. 5.) The rays are quite widely spaced with only a web of very flexible tissue connecting one to the next. This makes the fins more mobile in two ways. First, they can make undulations of relatively large amplitude and short wavelength. The dorsal fin of the sea horse, *Hippocampus*, has only about 19 segments, but it can be thrown into $3\frac{1}{2}$ waves.[26] When one ray is pointing directly dorsally, the next may be inclined as much as 17° laterally. Secondly, the fins of holosteans and teleosts

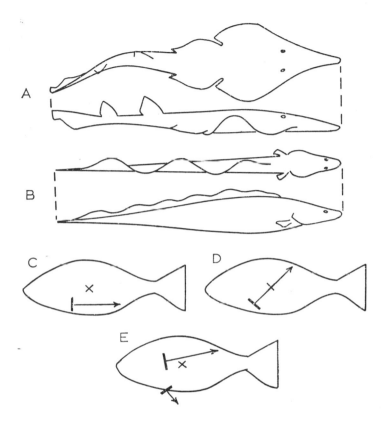

Fig. 7 Sketches and diagrams illustrating functions of fins. (A) Dorsal and lateral views of *Rhinobatos* undulating its pectoral fins to aid fast swimming. (B) Dorsal and lateral views of *Gymnarchus* swimming by undulating its dorsal fin. (C), (D), (E) Diagrams of teleosts braking showing the forces acting on the pectoral and, in (E), pelvic fins. The centres of gravity of the fish are marked X.

can be opened and closed like fans. Differentiation of muscle function has occurred to control the opening and closing movements, and it has occurred differently in the various fins. The particular arrangement which will be described is found only in the dorsal and anal fins.

The two halves of a ray lie on either side of the most distal radial, which also consists of two halves tightly bound together (fig. 5). They are attached to it by ligaments so that there is a hinge joint between ray and radial: the ray can swing forward and back about a transverse axis. The joint between the distal radial and the next is also a hinge joint, but its axis is longitudinal so that it allows the ray and distal radial to swing from side to side. Transverse movements involve the joint between the most distal radial and the next one, as in selachians. The rays of teleosts are thus attached to the body by two hinge joints with their axes at right angles. The arrangement is essentially similar to the universal joint used in the transmission system of cars. It enables the rays to swing forward and back or from side to side; in engineering terms, it allows each ray two degrees of freedom of movement relative to the body.

Animals usually have at least one pair of antagonistic muscles for each degree of freedom of relative movement in the body. Thus a hinge joint allows one degree of freedom and has at least two muscles while a ball-and-socket joint allows three degrees of freedom (rotation about three axes mutually at right angles) and has at least six muscles. The joint we are discussing allows two degrees of freedom and might be expected to have four muscles, but usually has six, as shown in fig. 5. The erector muscles swing the ray forwards, the depressors swing it back and the inclinators, contracting one at a time, swing it to one side or the other. Catfish (Siluriformes) have no inclinator muscles in the dorsal fin but can still swing the rays laterally.[10] Presumably they do this by contracting the erector and depressor of one side simultaneously. Even the four muscles present in catfish are more than the theoretical minimum. The joint between a ray and the body could be operated by three muscles, with each pair acting together as the antagonist of the third.

Most teleosts have swimbladders which make their density about equal to that of the water in which they live (Chapter 3). They do not need the lift which is provided in selachians by the pectoral fins and heterocercal tail. They usually swim with the pectoral fins laid flat against the body where they generate least drag. The possibility of putting them in this position is one of the consequences of their great mobility. Some teleosts, however, lack swimbladders and are denser than water. Among them, some Scombridae swim with their pectoral fins extended, presumably to give lift. The caudal fins of typical teleosts are symmetrical, lying posterior to the vertebral column instead of mainly ventral to it, and so do not produce an upward force when they move from side to side. This type of tail is called homocercal.

The mobile fins of holosteans and teleosts have a variety of functions. One of the most important is related to the possession of a swimbladder and the consequent ability of the fish to rest almost motionless in mid-water. A swimbladder with a given quantity of gas in it only gives the fish exactly the same density as the water at one particular depth, because the volume of the gas depends on the pressure (Chapter 3). Even at that depth the equilibrium of the fish is unstable. Any upward displacement makes the fish more buoyant so that it tends to rise further, and any downward one makes it denser so that it tends to sink further. All vertical displacements must be corrected by fin movements if the fish is to maintain its depth. The equilibrium is unstable in another respect. Because of the position of the swimbladder in the body cavity the centre of buoyancy is usually ventral to the centre of gravity.[70,77] If the fish leans to one side, the weight acting at the centre of gravity and the upthrust acting at the centre of buoyancy constitute a couple tending to turn the fish upside down. The fish can only stay the right way up by using its fins to correct accidental rolling movements. Fin movements are also needed because the respiratory current of a fish has a jet-propulsion effect, tending to drive the fish forwards. If the fish is to stay in one place, an equal and opposite force must be produced. Thus, although a fish with a swimbladder can hover in mid-water with very little expenditure of energy, quite a variety of small forces are needed to enable it to maintain its position. These forces are produced by passing waves along the fins. Many teleosts and the holostean *Lepisosteus* can often be seen in aquaria, hanging in mid-water, gently undulating various fins. The pectoral, caudal and dorsal fins are the ones most often used.

The effect of changes of depth has been simulated in experiments with perch (*Perca*), by subjecting them to reduced pressure.[61] When the pressure is reduced, the swimbladder expands and they become less dense than water. At first they maintain their position by means of their fins alone. The fins move faster as the pressure falls and are apparently unable to counteract the excess buoyancy of the fish when the upward force due to excess buoyancy is more than about 1% of the weight of the fish. This upward force is much less than the downward force of about 5%

of the body weight which is supported by a swimming dogfish. When the upward force is more than 1% of its weight, the perch can only avoid being carried to the surface by swimming downwards, using the whole body instead of just the fins.

Waves passed along fins can be used to propel a fish as well as to keep it in one position. Many teleosts use them for small movements. They allow a much greater range of movements than ordinary swimming with the tail. For instance, a fish can swim backwards by passing waves forwards along its fins; it can make pitching movements by passing waves up or down its caudal fin; and it can swim vertically up and down with its body horizontal by means of waves in the pectoral and caudal fins. Many acanthopterygians seem to be particularly adept at this type of movement. They tend to have larger fins than more primitive teleosts which use fin undulation mainly for holding their position. The acanthopterygian *Tilapia*, which uses its fins a good deal for manœuvring, can be seen in fig. 1 to have considerably larger fins than the primitive teleost *Salmo*, even if the spiny anterior part of the dorsal fin (which cannot be undulated) is ignored. Many of the fish which live on coral reefs are acanthopterygians with large fins, which make great use of fin movements as they manœuvre and search for food among the branches of coral.[5]

In some teleosts, fin undulation is the main mechanism of locomotion. The sea horse, for instance, swims by means of its dorsal and pectoral fins, and *Gymnarchus* (fig. 7B) and Gymnotoidei by means of their very long dorsal and anal fins, respectively. In *Gymnarchus*, at least, this type of locomotion seems to be reasonably efficient. We have seen how the power required for ordinary swimming by undulation of the whole body can be expected to be least when the waves pass backwards along the body little faster than the body moves forward through the water. The same should be true when only a fin is undulated. A sequence of stills from a film of *Gymnarchus* swimming forward at 0·5 lengths/sec shows waves travelling backwards along the dorsal fin at only 0·63 lengths/sec.[68]

Harris[49] has described how fins can produce forces in different directions, according to the manner in which they are moved. Symmetrical undulation produces a force in the plane of the fin, but asymmetrical undulation in which each ray moves faster in one direction than in the other may produce a force almost at right angles to it.

Another important function of the mobile fins of teleosts is as brakes. A fin laid flat against the body must add very little to the drag on a fish, but when it is extended from the body at a large angle to the direction of movement it can produce a very large pressure drag which will decelerate the fish. Quite a small fin can increase the drag on a fish very substantially. A thin flat plate moving through a fluid along an axis at right angles to its surface suffers many times as much drag as a streamlined body whose maximum transverse section has the same area as the plate.

The pectoral fins seem to be particularly important in braking, since they can be spread more or less at right angles to the direction of motion. In primitive teleosts such as *Salmo* (fig. 1) they are placed low on the body, ventral to the centre of gravity. When the pectoral fins are in this position, a fish which brakes by means of them alone must either make a pitching movement or else rise bodily in the water as it stops. If the pectoral fins are spread vertically so that the force on them consists entirely of horizontal drag (fig. 7C) it will pitch, with its head sinking and its tail rising. If they are held at such an angle that they produce lift as well as drag, and the resultant force on them acts through the centre of gravity of the fish (fig. 7D) there will be no tendency to pitch, but the fish will rise in the water as it stops. In some circumstances it may be desirable for a fish either to pitch or to rise as it stops. In others it will not. A fish cannot stop without pitching or rising unless other fins are used with the pectorals, or the pectorals lie on the same level as the centre of gravity.

The goldfish (*Carassius*) has its pectoral fins set low on the body. A film taken by Bainbridge[17] shows a goldfish braking. The pectoral fins are extended more or less at right angles to the body, the dorsal fin is thrown into a curve, and the caudal fin is bent laterally. All these fins must be helping to brake the fish. The drag on the dorsal fin, acting dorsal to the centre of gravity, probably counteracts the tendency of the drag on the pectoral fins to make the fish pitch.

Acanthopterygians have their pectoral fins set high on the body, roughly level with the centre of gravity (see *Perca* and *Tilapia* in fig. 1). If they held them vertically, they could presumably stop without either rising or pitching. However, the pectoral fins are held at an angle and apparently produce some lift. If the pelvic fins of the acanthopterygian *Lepomis* are amputated, it rises

whenever it brakes.[49] It seems that level stopping is only achieved
by a balance between the forces on the pectoral and pelvic fins,
which are used together in braking (fig. 7E). This balance is only
possible because of the position of the pelvic fins. Instead of being
posterior to the centre of gravity, as in primitive teleosts, they
are more or less immediately ventral to the pectoral fins.

When a teleost is braking it is desirable that every fin should
produce as much drag as possible, but when it is swimming they
should produce as little drag as possible. When it is held parallel
to the direction of motion a fin produces much less drag than
when it is held at a large angle, but it still produces some drag. It
seems that this drag can be reduced by folding the fin. Teleosts
often fold their fins when they start swimming, and erect them as
they stop. The *Gadus* shown in fig. 1 is swimming and has its
fins partly folded, whereas the *Salmo*, *Perca* and *Tilapia* are hovering and have their fins extended. Only the pectoral fins can be laid
flat against the body, but Scombridae have a groove which houses
the anterior dorsal fin when it is depressed.

The Acanthopterygii

The Acanthopterygii are the largest superorder of fish. We have
already noted the great use that typical acanthopterygians make
of fin movements for manœuvring. We have also noted the
positions of their pectoral and pelvic fins and the way they act as
brakes. We will now consider their spiny fin rays and their shape.

Acanthopterygii owe their name to their spiny fin rays. These
are solid pieces of bone, whereas an ordinary ray consists of a
large number of tiny pieces of bone. They are therefore much
stiffer than ordinary rays. They are not branched, and are pointed
at their ends. The radials of each spiny ray are fused or sutured
together, so that the distal one is no longer movable on the more
proximal ones. The spiny rays therefore cannot move from side
to side, though they can be erected and depressed in the usual
way.

Spiny rays are found at the anterior ends of the pelvic, anal
and dorsal fins (fig. 1, *Perca* and *Tilapia*). They are usually most
numerous in the dorsal fin, which is often divided into an anterior
spiny fin and a posterior soft-rayed one. The spiny fin lies largely
anterior to the centre of gravity, and therefore tends to make the
fish unstable in yaw when it is erected. This effect is enhanced

by the stiffness of the rays and by their being unable to swing laterally. Harris[49] has described how acanthopterygians often extend one pectoral fin and erect the spiny part of the dorsal fin when they turn. The pectoral acts as a brake on its side of the body and starts the turn, while the instability due to the erected spiny dorsal helps its completion. More primitive teleosts turn simply by bending their bodies to one side.

This may be an important function of the spiny rays, but it cannot entirely explain them. It does not explain the continuation of the spiny part of the dorsal fin posterior to the centre of gravity in fish like *Tilapia*. It does not explain the spines in the pelvic and anal fins. The spines have probably evolved largely as a defence against predators. Defensive fin spines will be discussed further in Chapter 7.

Acanthopterygians often have rather short, deep bodies. *Tilapia* (fig. 1) is typical. This shape can be related to the great manœuvrability they enjoy through the use they make of their fins. When a fish rotates its body about an axis, the parts of the body which are far from the axis move faster than the parts which are close to it. The faster a part of the body moves, the greater the drag resisting its motion and the greater the power used in overcoming the drag. If a fish uses its fins to rotate its body about a transverse axis in a pitching movement, or about a vertical axis in a yawing movement, the power required for a given angular velocity is least if the body is short. Similarly, if a fish turns by braking with one pectoral fin, the amount of momentum it loses is least if its body is short. Consequently typical acanthopterygians have become relatively short by evolving deep, compressed bodies (fig. 3G).
sequently typical acanthopterygians have become relatively short by evolving deep, compressed bodies (fig. 3G).

The deep bodies and large fins which make typical acanthopterygians so manœuvrable lead to increased drag in forward swimming. They are advantageous to fish in habitats where manœuvrability is particularly important, such as among vegetation in lakes, or in coral reefs. The acanthopterygians are particularly common in such habitats. They would tend to be disadvantageous to fish which swam continuously in open water, as the Scombridae do. Although the Scombridae are acanthopterygians, they are only slightly compressed (fig. 3E) and have relatively small median fins, though some have long pectoral fins.

3

BUOYANCY

A fish which has no special buoyancy organ usually has a specific gravity between 1·06 and 1·09 and is therefore denser than either fresh water (specific gravity 1) or sea water (1·026). This is simply due to the materials of which it is made: the specific gravity of fish muscle is usually around 1·05, of cartilage 1·1, of bone and bony scales 2·0 and of fats 0·9. Of these, only fats are less dense than water, and as they are only a little less dense than water it takes a lot of fat to make a small difference to the density of a fish.

A fish which is denser than the water in which it lives is said to have negative buoyancy, and this is the condition of fish without special buoyancy organs. One which is less dense than the water has positive buoyancy, and one which is exactly the same density as the water has neutral buoyancy. Neutral buoyancy has, as we shall see, certain advantages, and many fish have approximately neutral buoyancy. They have achieved this in two principal ways. Some sharks have achieved it by evolving huge livers which consist largely of a hydrocarbon of specific gravity 0·86.[25a] This hydrocarbon may form as much as 20% of the total weight of the body. Many teleosts achieve neutral buoyancy by means of a swimbladder which is a gas-filled sac in the body cavity.[11,38] Quite a small swimbladder can make a big difference to the density of a fish, because the gas increases the fish's volume without appreciably increasing its weight. A typical fish without a swimbladder would be about 5% denser than sea water, and could acquire neutral buoyancy in sea water by evolving a swimbladder whose volume was 5% of the initial volume of its body. It would need a slightly bigger swimbladder, in fresh water. This chapter is mainly about swimbladders.

Swimbladders seem to have evolved from lungs. Among modern bony fish, the dipnoans, holosteans and Brachiopterygii have either a single lung or a pair of lungs. The crossopterygians from which the tetrapods evolved presumably had lungs. It seems likely that the common ancestor of the bony fish had a lung and breathed air, and that the typical teleost swimbladder is a lung which has lost its respiratory function. A lung needs a trachea, and the more primitive teleosts retain a duct between the gut and the swimbladder, but this has been lost by the more advanced teleosts, including the acanthopterygians and paracanthopterygians. An air-filled lung necessarily makes a fish more buoyant, but the lungs of the early bony fish were probably not big enough to give them neutral buoyancy. The heterocercal tails of these fish (Chapter 2) strongly suggest that they had negative buoyancy, though one must be rather cautious in inferring the density, in life, of a fossil from the shape of its tail. Many teleosts have negative buoyancy and symmetrical tails, while *Lepisosteus* (fig. 2) retains an appreciably heterocercal tail although it has neutral buoyancy. The early bony fish would have needed very big lungs for neutral buoyancy, as they had very thick bony scales.

Advantages of neutral buoyancy

A fish without a buoyancy organ is usually about 5% denser than sea water. When it is in the sea it can only keep off the bottom by generating an upward force equal to about 5% of its own weight. A rather bigger force is needed in fresh water. No fish seems able to produce such forces while it is stationary. Fish without buoyancy organs cannot hover in mid-water; they can only keep off the bottom by swimming around, using the pectoral fins, a heterocercal tail or even the whole body as hydrofoils to give the necessary lift (Chapter 2). They can only rest by allowing themselves to sink to the bottom. This may be no disadvantage to fish like the flatfish (Pleuronectiformes), which find their food on and near the bottom and spend much of their time resting on the bottom, but it might be a serious disadvantage to a fish which found its food near the surface in deep water. Fish with swim-bladders which give them neutral buoyancy can and often do hover

in mid-water. They still have to make gentle fin movements, because their equilibrium is unstable (Chapter 2). Aquarium fish
which hover by day often rest on the bottom by night, and divers
have seen American perch (*Perca*) behaving in the same way in a
lake. The habit of resting on the bottom at night may have evolved
to save the small amount of energy needed for fin movements
when hovering, or it may be related to the mysterious process
of sleep.

As well as being unable to hover a fish without a buoyancy
organ needs more power to swim at a given speed than one with
neutral buoyancy. This is because lift cannot be obtained without
drag. In Chapter 2 we divided the drag on a fish into friction drag
and pressure drag. In this chapter it will be convenient to make
a different division which cuts across the previous one. We will
think of the drag as consisting of zero-lift drag, which is the
minimum value to which the drag could be reduced if no upward
lift were required, and drag due to lift. If the upward lift is provided by fins, the drag due to lift will be approximately equal to the
drag on the fins which provide the upward lift, and the zero-lift
drag to the drag on the rest of the body. We will estimate the
power required, by a fish without a buoyancy organ, to overcome
drag due to lift.[11]

We will suppose that the fish is 5% denser than the water in
which it lives. If it weighs W gm, the upward lift which it needs
will be about $0.05W$ gm wgt or $50W$ dynes. The lift-drag ratio of
the fins which provide the lift might be about 8; this is a guess
based on the properties of locust wings, which work at Reynolds
numbers near those of many fish fins.[14a] If so, the drag due to lift
will be about one-eighth of the upward lift, or $6W$ dynes. The
power required to overcome it at a swimming speed of Vcm/sec
will then be $6WV$ ergs/sec. The metabolic rate required to produce
this power will depend on the efficiency of the muscles which is
unlikely to be much more than 20%, and on the efficiency of
swimming. The relationship between oxygen consumption and
power was explained in Chapter 2. If we allow for 10% overall
efficiency, we can estimate that the part of the metabolic rate
which provides the power needed to overcome drag due to lift
is $1.4V$ mgO$_2$/Kg body wgt hr.

We will apply this estimate to the 16.5 cm *Onchorhynchus*
whose metabolic rates at various swimming speeds are shown in
fig. 4. If it had no swimbladder it would have to increase its

metabolic rate by about $1.4 \times 16.5 = 23$ mgO$_2$/Kg hr when swimming at 1 length/sec, by 46 mgO$_2$/Kg hr at 2 lengths/sec and by 92 mgO$_2$/Kg hr at 4 lengths/sec. These figures are 15%, 21% and 12%, respectively, of the actual metabolic rates at the same speeds.

One factor has been overlooked so far. A buoyancy organ increases the size of a fish, and so the zero-lift drag. If the fish is initially 5% denser than the water, it can obtain neutral buoyancy with a swimbladder which increases its volume by 5%. Friction drag and pressure drag on bodies of the same shape but different sizes are roughly proportional to surface area. A 5% increase in volume will tend to cause an increase of about 3% in surface area and zero-lift drag. This will increase the power needed to overcome zero-lift drag by 3%, and so increase the metabolic rate by amounts which are very low at low speeds and approach 3% at high speeds. (Fig. 4 indicates that the energy required for swimming accounts for most of the metabolic rate at high speeds.)

To estimate the effect of a swimbladder on the metabolic rate, we must subtract the increase due to increased zero-lift drag from the decrease due to upward lift not being needed. If *Onchorhynchus* had no swimbladder, its metabolic rate might be expected to be about 15% higher when it was swimming at 1 length/sec, 20% higher at 2 lengths/sec and 9% higher at 4 lengths/sec. A swimbladder probably allows a substantial saving of energy in swimming. The hydrocarbon-filled livers of pelagic sharks probably save rather less energy, because the volume of hydrocarbon needed for neutral buoyancy is larger than the volume of gas needed in a swimbladder, and the increase in zero-lift drag must be correspondingly more.

When mackerel and other Scombridae are kept in aquaria they never stop swimming. They presumably swim all the time in the sea. Nevertheless, many of them have no swimbladder and are denser than sea water.[11] This suggests that the possession of a swimbladder may have some disadvantage to them which can offset the advantage due to the energy it saves. A possible disadvantage has been suggested to me by Dr D. W. Yalden. Swimbladders are good reflectors of underwater sound. Toothed whales practise echo-location, so it may be easier for them to find a fish with a swimbladder than one without. Dolphins eat mackerel, among other fish,[3] and killer whales eat tunnies.

Negative buoyancy has an important advantage for fish which rest on the bottom. When a fish with negative buoyancy rests on the bottom, friction tends to hold it in place. A fish with neutral buoyancy has no weight in water, and so no friction. Many teleosts which live and feed on the bottom, especially ones which live in rivers and on shores, have reduced swimbladders or no swimbladder at all. Apparently the advantages of negative buoyancy can outweigh its disadvantages in these conditions. A river fish must not allow itself to be washed passively downstream if it is to stay in its habitat, and it may be able to maintain its position more economically by exploiting negative buoyancy and resting on the bottom than it could with neutral buoyancy swimming perpetually against the current. If it has negative buoyancy, swimming will need more energy while it lasts, but will not have to be continuous. It may be no disadvantage to a shore fish if it is carried to and fro by tidal currents, but a fish which rests on the bottom, particularly in sheltered crannies, will be less liable to be damaged by waves than one which swims all the time.

Among common British river fish, the stone loach, *Nemacheilus*, has a reduced swimbladder, and the miller's thumb, *Cottus*, has none. Quite a lot of others have, rather surprisingly, well-developed swimbladders which give them roughly neutral buoyancy. An example is the dace, *Leuciscus*, which lives mainly in streams and swims, apparently continuously, against the current. In one stream I know, 15–25 cm dace keep station against a current of 17–30 cm/sec. They must use a good deal more energy than if they rested on the bottom. Dace feed largely on insect larvae taken from the bottom.[50] Perhaps they can catch more larvae swimming above the bottom than they could if they crept about on the bottom. If the habit of perpetual swimming increases a fish's success in finding food more than enough to pay the metabolic cost of swimming, it will have a selective advantage.

Many fish which live in really fast mountain streams have not only got negative buoyancy, but have suckers as well. In *Gyrinocheilus* the lips are expanded to form a relatively small sucker round the mouth, while in *Gastromyzon* the whole of the broad ventral surface of the body acts as a sucker. Both these examples are Cyprinoidei living in the mountain streams of S. Asia, but suckers have also evolved in other groups and continents. A fish attached by a sucker still depends on friction to resist horizontal forces, but the extra vertical force applied by the sucker increases the friction. Most torrent-living fish seem to feed on algae which encrust the rocks. They can collect quite a lot of food from one rock before making a strenuous dash for another. Many shore fish such as the lumpsuckers (*Cyclopterus*) have a sucker as well as having no swimbladder. They use the sucker to hang onto rocks, and it must help to protect them from being thrown about and battered by wave action.[40a]

Buoyancy and depth

As a buoyancy organ, the swimbladder has a serious disadvantage. Its volume depends on the pressure, and so on the depth of the fish. There is only one depth at which a fish has neutral buoyancy, and even at that depth its equilibrium is unstable (Chapter 2). We shall see presently that the quantity of gas in the swimbladder can be altered to change the depth of neutral buoyancy. We shall also see that the process of adjustment tends to be slow.

When teleosts hover in mid-water they use their fins to counteract small vertical forces due to positive or negative buoyancy (Chapter 2). They can only do this when they are reasonably near the depth of neutral buoyancy; experiments in which *Perca* were subjected to reduced pressures indicated that a perch which had neutral buoyancy at a depth of 2 m would only just be able to maintain its depth by fin action as it approached the surface, and one with neutral buoyancy at 20 m would only just be able to do so at 15 m.[61] Negative buoyancy due to greater reductions of pressure can be resisted by swimming downwards, using ordinary swimming movements involving the whole body, but even this has a limit. The experiments indicated that a perch with neutral buoyancy at 20 m would be carried helpless to the surface if it swam up to a depth of 10 m or less, but the experimental fish were in rather a small container and might have done better if they had had more room to swim.[11] Whatever the limit to upward movement is, it seems plain that it exists. Greater upward movements require removal of gas from the swimbladder. There is probably no corresponding lower limit to the vertical range of a fish below which it will be carried helpless to the bottom, since no matter how great the pressure is, the fish will never be denser than if it had no swimbladder at all. Plenty of teleosts lack swimbladders but can still swim. Herring (*Clupea*) which had neutral buoyancy at the surface have been lowered in a cage to 30 m and observed by underwater television; they swam normally. However, a species which made a habit of hovering would probably, as it descended, reach a depth at which it could no longer hold its position by fin movements.

The change of buoyancy with depth, due to compression and expansion of the swimbladder, is plainly disadvantageous. The smaller the rate of change of buoyancy with depth, the less the energy the fish has to use when it hovers above or below the depth of neutral buoyancy. The rate will be small if the swimbladder is small, so there is an advantage in keeping the specific gravity of the tissues low. This may be why the scales of bony fish became thinner in the course of evolution. The rate is kept low in a different way in Cypriniformes. Their swimbladders have rather inextensible walls, inflated with gas under pressure. Such swimbladders expand less than ordinary ones at reduced pressures, and are compressed less at increased pressures. The volume of the swimbladder of the bream (*Abramis*) changes only one-quarter as much, for small changes of depth, as would the volume of an ordinary swimbladder.[11]

The more primitive teleosts, and some advanced ones including most Ostariophysi, have a duct from the swimbladder to the gut. Air gulped at the surface can be admitted through it, and swimbladder gases can be expelled through it, just as air can be breathed in and out of a lung. When fish with swimbladder ducts are subjected to increased pressures in laboratory experiments, they gulp air at the surface to restore their buoyancy, but this process seems unlikely to be much use in nature. A fish intending to swim at a certain depth would have to visit the surface, gulp enough air to give it neutral buoyancy at the depth in question, and then swim down to that depth (if the resulting positive buoyancy at the surface was not so great as to prevent downward swimming). Various teleosts with swimbladder ducts are known to be able to secrete gases into their swimbladders and presumably use this ability. In experiments in which the pressure is reduced, teleosts with swimbladder ducts release gas through them. They probably do the same when they swim from deep water towards the surface. Teleosts without swimbladder ducts can only alter the quantity of gas in the swimbladder by secretion from the blood or resorption into it.

Partial pressures

When we come to discuss the secretion and resorption of gases, it will be convenient to use the concept of partial pressure. The partial pressure of a gas in a mixture of gases is the pressure that it would exert if it alone occupied the whole volume of the mixture. Air contains about 20% oxygen by volume, so the partial pressure of oxygen, in air at atmospheric pressure, is 0·2 atm. When a mixture of gases is in equilibrium with a liquid, the amount of each gas in solution in the liquid is proportional to its solubility, and to its partial pressure in the gaseous phase. The partial pressure of a gas in solution is defined as the partial pressure it would have in a mixture of gases which was in equilibrium with the solution. Thus, the partial pressure of dissolved oxygen in water which is in equilibrium with the atmosphere is 0·2 atm.

The gas in the swimbladders of fish in shallow water often has about the same composition of air. In experiments when the swimbladder is emptied and newly secreted gases are collected, much higher proportions of oxygen are found. Once it has been secreted the oxygen is gradually replaced by nitrogen until the mixture, like air, contains about 0·8 atm nitrogen and only 0·2 atm oxygen. Fish at moderate depths also tend to have 0·8 atm nitrogen in their swimbladders. At a depth where the pressure is P atm they therefore tend to have $(P-0·8)$ atm oxygen, for the total of the partial pressures of the gases in a mixture must equal the pressure of the mixture. Thus the proportion of oxygen in the swimbladder increases with depth. At great depths, swimbladders usually contain more than 0·8 atm nitrogen. The swimbladder gases of several species caught at 1,200 m have been found to contain about 8% nitrogen and 92% oxygen. The pressure in water increases by 1 atm for every 10 m increase in depth, so the pressure at 1,200 m is 121 atm. The partial pressure of the nitrogen at this depth must have been 8% of this or about 10 atm.

The Salmonidae have exceptionally high proportions of nitrogen in their swimbladders. Even newly secreted gases may be almost pure nitrogen. Salmonidae with 95% or more nitrogen in their swimbladders have been caught in lakes in depths down to about 150 m, where the partial pressure of the nitrogen would be 14 atm or more.[11]

Resorption

The resorption of gas from the swimbladder is essentially the same process as the absorption of oxygen from a lung. The water in which fish live normally gets its dissolved gases from the atmosphere. It therefore normally contains 0·8 atm nitrogen and 0·2 atm or less oxygen, though higher partial pressures of oxygen may be built up by photosynthesis where there are a lot of water plants. The blood of fish tends to come into equilibrium with the water at the gills, and so to contain 0·8 atm nitrogen and 0·2 atm or less oxygen. When, as is usual, the swimbladder contains about 0·8 atm nitrogen and $(P-0·8)$ atm oxygen, the nitrogen is more or less in equilibrium with the blood. The oxygen, however, will tend to dissolve in the blood whenever it is exposed to it, since $(P-0·8)$ is greater than 0·2 so long as the fish is submerged.

There has therefore been no need to evolve a mechanism for resorbing gases from the swimbladder. It has, rather, been necessary to evolve a means whereby resorption could be controlled and more or less halted, by controlling access of the gases to the blood. Lungs need a rich blood supply so that their oxygen can diffuse quickly into the blood. In the evolution of teleost swimbladders most of the blood vessels have become localised in specialised organs of gas secretion and resorption.

The organ of resorption consists merely of a region of the swimbladder wall which has plenty of blood vessels. In the eel (*Anguilla*) it is a modified swimbladder duct. Gas can be resorbed into the blood in the vascular wall of the duct, or simply expelled through the duct. Normally, the duct is slender and little blood flows through its vessels. When the eel is resorbing gas, however, the duct swells and fills with gas from the main cavity of the swimbladder. Its blood vessels dilate, allowing far more blood to flow through them. Experiments in which rates of resorption have been measured indicate that resorption is entirely due to passive diffusion of gases into the blood. Oxygen can even be made to pass into the swimbladder from the blood vessels of the duct, by filling the swimbladder with nitrogen.[87] Myctophidae, Acanthopterygii and Paracanthopterygii have no swimbladder duct, and à posterior part of the swimbladder wall is specialised as a resorbent organ. Sometimes it is a pocket in the swimbladder wall which can be closed by a sphincter when resorption is not required. This type of organ (the oval) is represented in fig. 8A.

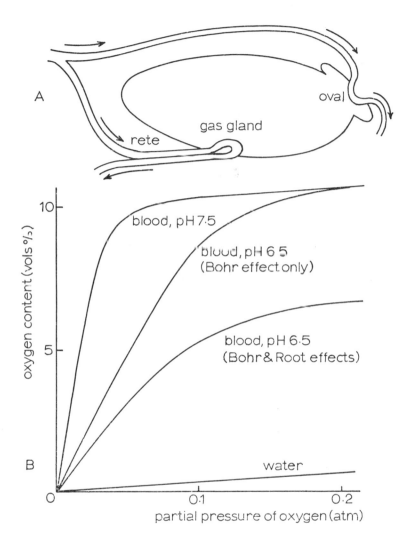

Fig. 8   (A) A diagram showing the blood supply of a swimbladder and (B) graphs of oxygen content against partial pressure of oxygen for typical teleost blood and for solutions of oxygen in water. The graphs are based on data for many species given in ref. 1

Sometimes the resorbent organ is a chamber which cannot be isolated, and resorption is controlled by contraction of muscles which reduce its area and thicken its epithelium. In both cases, the flow of blood can be controlled by constriction of the blood vessels.[38]

We are going to work out the rates at which teleosts are likely to be able to resorb gas from their swimbladders. As the nitrogen in the swimbladder is usually more or less in equilibrium with the blood, only oxygen need be considered. It will be inconvenient to describe quantities of gas in terms of volume, because we will be considering different depths and so different pressures. It will be better to use moles. One mole of gas in the range of temperatures at which fish live occupies about 23 litres at atmospheric pressure. 1 ml of gas, measured at atmospheric pressure, is therefore 1/23 mM.

Oxygen can be removed from the swimbladder as oxyhaemoglobin or in physical solution in the blood. The oxygen capacity of the haemoglobin of teleost blood is usually about 10 ml oxygen at atmospheric pressure/100 ml blood, or 0·4 mM oxygen/100 ml blood (except in the presence of acid or a high concentration of carbon dioxide).[1] The haemoglobin becomes more or less saturated at low partial pressures of oxygen (often about 0·05 atm, see fig. 8B), and partial pressures higher than this force hardly any more oxygen into combination with it. The haemoglobin of the arterial blood arriving at the swimbladder is likely to be about 85% saturated,[1] in which case every 100 ml of the blood will be able to take up 0·06 mM oxygen as oxyhaemoglobin. The physical solubility of oxygen in blood is about 4 ml/100 ml blood or 0·17 mM/100 ml blood atm. At a pressure of P atm, the difference in partial pressure of oxygen between the swimbladder and the blood is usually about $(P-1)$ atm. 100 ml blood will therefore be able to take up 0·17$(P-1)$ mM oxygen in physical solution. It can therefore remove, in total, $0·06 + 0·17(P-1) = (0·17P - 0·11)$ mM oxygen from the swimbladder.

The rate at which blood is pumped through the heart of a teleost is probably usually about 60 ml/100 gm body wgt hr.[59] Of this, as much as 20% or 12 ml/100 gm hr may go to the resorbent part of the swimbladder when resorption is in progress, at least in eels.[87] This will be able to remove $(0·17P-0·11)12/100 = (0·02P - 0·013)$ mM oxygen/100 gm body wgt hr from the swimbladder. The volume of the swimbladder of a teleost which has neutral buoyancy is typically about 5 ml/100 gm body wgt. At a depth where the pressure is P atm it will contain about $0.2P$ mM gas/100 gm body wgt. To reduce its depth of neutral buoyancy by 1 m, the fish must remove 0.02 mM gas/100 gm body weight, since a rise of 1 m reduces P by 0.1. This will take $0.02/(0.02P - 0.013)$ hr. Hence, the maximum rate at which a teleost can reduce its depth of neutral buoyancy by resorbing gas from the swimbladder seems likely to be about 0.3 m/hr near the surface where $P = 1$, and about 10 m/hr at a depth of 100 m where $P = 11$. The first of these estimates can be checked because rates of resorption in shallow water have been determined experimentally. *Perca*, for instance, which admittedly have rather larger swimbladders than were assumed in the calculation, take about 9 hr to adapt when brought from 3 m to the surface.[61] The rate of resorption must be highly dependent on the degree of saturation of the haemoglobin near the surface, but more or less independent of it at considerable depths.

This calculation seems to show that a fish which depends on a resorbent organ to remove gas from its swimbladder is most unfavourably placed in comparison with one which has a swimbladder duct. The rate at which it can move upwards is severely limited, except at great depths. A fish with a duct can presumably move upwards as fast as it can swim, spitting out gas as required. The loss of the swimbladder duct by the more advanced teleosts is very puzzling. There is no respiratory advantage in resorption into the blood, since the blood carrying the resorbed oxygen returns directly to the heart and is equilibrated with the water at the gills.

Secretion

Resorption of gases from the swimbladder is a passive process and, for the same reason, secretion must be active. Teleosts can apparently secrete gases against very large differences of partial pressure. Fish with swimbladders have been caught at depths down to 7,000 m, where the total partial pressure of dissolved gases in the water is no more than at the surface, but the pressure of the gas in the swimbladder must be 700 atm.

Some lower teleosts have no distinct organ for gas secretion which probably occurs over large areas of the swimbladder wall. Other teleosts have well-defined glandular patches which are made conspicuously red by their rich blood supply. There is no

doubt that these serve to secrete gas, since bubbles can be collected from them when the swimbladder of a living fish is laid
open. They are known as gas glands.

The mechanism of gas secretion has only been 'demonstrated
in *Anguilla* which is a long way from the main line of teleost evolution, but there is no reason to believe that the mechanism is
different in other teleosts, except for the Salmonidae. It depends
on the secretion of lactic acid at the gas gland. During gas
secretion, the blood leaving the gas gland contains more lactic
acid than the blood arriving at it.[88]

To release gases from the blood, it is necessary to raise their
partial pressures till they are greater than the partial pressures
in the swimbladder. There are several ways in which lactic acid
can increase the partial pressures of gases in the blood by small
amounts.[11] When we have considered them we will go on to see
how the small increases are multiplied to give the enormous
increases which are needed for secretion at great depths.

Acid tends to dissociate oxyhaemoglobin (fig. 8B). In vertebrates in general, a higher partial pressure of oxygen is needed to
saturate the haemoglobin in the presence of acid. This effect (the
Bohr effect) is important in respiration because, through it,
carbon dioxide and lactic acid accumulating in active tissues
speed the release of oxygen from the blood. It is particularly
marked in most teleosts, but we will note some exceptions in
Chapter 4. In many teleosts acid does not merely increase the
partial pressure of oxygen required to saturate the haemoglobin,
but actually decreases the quantity of oxygen which can be taken
up, even at very high partial pressures (the Root effect). Acid also
tends to release carbon dioxide from bicarbonate in the blood. The
lactic acid secreted at the gas gland therefore increases the partial
pressures of oxygen and carbon dioxide in the blood.

It also increases the partial pressures of gases by the salting-
out effect. Ions reduce the solubility of gases; for instance oxygen is 20% less soluble in sea water than in fresh water. The salting-out effect is small, but it affects all the gases dissolved in the
blood. It is possible that the Salmonidae, which secrete nearly
pure nitrogen, secrete a mildly basic substance into the blood
instead of lactic acid. By doing so they could increase the partial
pressure of nitrogen without altering the partial pressure of
oxygen.

These mechanisms alone would not release gases at any substantial depth. If all the oxyhaemoglobin could be dissociated (which is most unlikely) it might yield 0·4 mM oxygen/100 ml blood. The physical solubility of oxygen in blood is about 0·17 mM/ 100 ml atm and is only reduced a little by the salting-out effect. Hence, any oxygen that was released from oxyhaemoglobin would go into physical solution in the blood at depths where the pressure was more than about 2·5 atm.

The increases in partial pressure produced (at least in the eel) by lactic acid are greatly magnified by the process of counter-current multiplication which has been explained by Kuhn and his colleagues.[65] This occurs in the rete mirabile through which the blood passes on its way to and from the gas gland (fig. 8A). In the rete, the artery and vein each break up into a large number of capillaries. The arterial and venous capillaries mingle and run parallel to each other. They are closely packed, and it is assumed that gases can diffuse rapidly between them. This dense mass of capillaries is represented in the figure by a single arterial and a single venous channel. Blood from the aorta travels to the gas gland in the arterial capillaries. At the gas gland, lactic acid is secreted into it, increasing the partial pressures of the gases. It then passes into the venous capillaries. There is now a difference of partial pressure between the gases in the blood in the two sets of capillaries, and gases diffuse from the venous to the arterial capillaries. The blood arriving at the gas gland thus comes to have an enhanced gas content. The partial pressures of its gases are, in turn, raised by secretion of lactic acid into it, and when it returns to the rete gas diffuses from it to the blood in the arterial capillaries. In this way, more and more gas accumulates at the gas gland end of the rete. High concentrations and partial pressures are built up. When the partial pressure of any gas comes to exceed its partial pressure in the swimbladder, it starts diffusing out of the gas gland into the swimbladder.

This process can be analysed mathematically. It can be shown that the rate of secretion of a gas should decrease as its partial pressure in the swimbladder increases, and that the maximum partial pressure against which secretion is possible should increase sharply as the length of the rete and the rate of secretion of lactic acid increase.[11, 65]

The concentration of lactic acid in the blood of secreting eels has been found to increase by about 5 mM/l as it passes through the gas gland.[88] This increase is similar in size to the increases of lactic acid concentration which occur in human blood during exercise. Estimates have been made of the maximum partial pressures against which it could cause secretion of gases, with retia of various lengths.[65] With a rete 1 cm long, secretion of oxygen would apparently be possible at partial pressures up to 2,000 atm (when there is a Root effect) or 300 atm (when there is not). Eels have retia up to 1 cm long but fish living at 2,000 m or more (200 atm or more) tend to have longer ones.[5] The observed rate of secretion of lactic acid should make secretion of nitrogen possible against a partial pressure of 6 atm when the rete is 1 cm long. A larger estimate which has been published depends on the supposition that salts may be secreted as well as lactic acid.[65]

Secretion, like resorption, is slow. Teleosts without swimbladder ducts, whose swimbladders are emptied in experiments, usually require 4–48 hrs to refill them. *Salmo* takes about a fortnight if it is prevented from gulping air at the surface. Even a fish which could fill its swimbladder in 4 hrs could not maintain neutral buoyancy in a rapid descent. If it needed x moles of gas for neutral buoyancy at the surface, it would need $2x$ at 10 m, $3x$ at 20 m. and so on. If it took 4 hrs to secrete every x moles, it would not be able to maintain neutral buoyancy in descents faster than 2·5 m/hr. The rate would be less at substantial depths, where secretion would be slower.[11]

Vertical movements

Many fish make regular periodic changes of depth.[14c] Seasonal changes, such as are made by *Perca* in lakes, seem to be accompanied by appropriate changes in the quantity of gas in the swimbladder. Daily changes, however, are often so rapid that it would seem impossible for the fish to secrete and resorb gases quickly enough to maintain neutral buoyancy.

Planktonic animals tend to swim towards the surface at dusk and away from it at dawn, so that their depth is less by night than by day. Various fish take part in these movements, feeding on smaller plankton near the surface at night and retiring to deeper (and so darker) water by day. The movements of the fish can often be followed by echo-sounding because fish swimbladders reflect sound well. When fishing and echo-sounding are done in the same place, it is often possible to decide what species of fish is mainly responsible for the echo-sounder traces.

Herring (*Clupea*) make vertical movements whose extent varies with the season and with the depth of the water.[24] In the North Sea they spend the night within 40 m of the surface and the day at about 150 m. Even the fastest secretors of gas, among the teleosts that have been investigated, could not compensate for the morning increase of depth within a day. Herring, however, seem to be very slow secretors of gas: indeed, laboratory experiments have failed to demonstrate any secretion at all. It seems very unlikely that herring can maintain neutral buoyancy during their daily movements. More probably they keep a constant quantity of gas in their swimbladders, which gives them neutral buoyancy somewhere near their night-time depths but leaves them negatively buoyant at their greater depths by day. The depth of neutral buoyancy could not be much below the night depth, for, as we have seen, a fish cannot swim far above its depth of neutral buoyancy without removing gas from the swimbladder. Herring kept in aquaria have roughly neutral buoyancy.

More spectacular daily movements are made by some smaller pelagic marine fish. The commonest of these are the Myctophidae or lantern fishes, which owe their common name to luminous organs on their skin. They have swimbladders of about the usual size for marine teleosts. Their movements were first inferred from the results of fishing at different depths and different times of day. Later, they came to be suspected of being one of the constituents of the deep scattering layers. These are layers of objects detectable by echo-sounding, which make daily movements in the oceans, up and down, through several hundred metres. Recently a deep scattering layer has been observed directly from a 'diving saucer', which is a submarine equipped with lights.[19] Myctophids and siphonophores (coelenterates with gas-filled floats) were plentiful at the depths indicated by echo-sounding. The myctophids were swimming in the top 100 m of the sea at 5 a.m., but soon began to swim down and by 8 a.m. had reached their daytime depth of around 300 m. They swam up more slowly in the evening. It seems most unlikely that they could maintain neutral buoyancy in these movements. No experiments have been done to find out how rapidly myctophids can secrete gas, but if they were to maintain neutral buoyancy in a descent of 200 m or so in 3 hrs, they would have to secrete at about 25 times the maximum rate that has been measured in experiments with other teleosts.

Deep scattering layers reflect certain frequencies of sound

particularly strongly. These are believed to be the resonant frequencies of the swimbladders of the fish and of the floats of the siphonophores. When the pressure and resonant frequency of a gas bubble is known, its size can be calculated. The manner in which the resonant frequencies of the deep scattering layers change as the layers descend at dawn and rise at dusk, indicates that in some layers the volumes are inversely proportional to the pressure, as would occur if the quantity of gas in the swimbladders or floats was constant. In at least one case, however, a layer rising in the evening seemed to be keeping the volume of its swimbladders or floats constant.[51]

Herrings and myctophids live mainly in the upper parts of the sea, though herrings may go close to the bottom by day in shallow water. Various other fish spend the day on or near the bottom, and swim upwards at night without coming close to the surface. Their movements are inferred from the results of trawling at different times of day, and from echo-soundings.[22] They include cod (*Gadus*) and hake (*Merluccius*). There is some evidence which suggests that such fish keep a constant quantity of gas in their swimbladders, which gives them approximately neutral buoyancy at their night depths. Scholander and his colleagues trawled various fish from depths of 20–90 m. The swimbladders of these fish were of course immensely swollen or burst when they reached the surface, owing to the reduction in pressure, but some species did not seem to have lost any swimbladder gas from the body cavity. The pressures at which these species had neutral buoyancy were determined. Two such species, which are believed to spend all their time near the bottom, had neutral buoyancy at pressures corresponding to the depths at which they were caught. Three others which were believed to swim up from the bottom at night were neutrally buoyant at much lower pressures which may have corresponded roughly with their night depths. The correspondence cannot have been exact, for in several cases the pressure was less than atmospheric pressure.[11]

We know a good deal less about the use teleosts make of their ability to secrete and resorb swimbladder gases than we do about the physiology of these processes. Most of the evidence we have indicates that teleosts that make daily changes of depth do not adjust the quantity of gas in their swimbladders accordingly, but keep a constant quantity of gas in the swimbladder which gives them neutral buoyancy at the top of their vertical range. This does not mean that the ability to secrete gas is used only to compensate for long-term changes of depth, and for growth of the fish. The swimbladder wall cannot be utterly impermeable to gases, and there must always be some diffusion of gases out of it. This loss must be made good by secretion. The rate at which it occurs in conger eels (*Conger*) has been investigated[35a]: it seems to be trivial in shallow water but would be substantial at depths of 1000m or more.[14c]

Deep-sea fishes

Many fish which swim in mid-water at depths of 500 m or more have rudimentary swimbladders, or no swimbladder at all. Two such species which have been investigated are only about 1% denser than sea water, whereas most teleosts without swimbladders are about 5% denser than sea water. This is because they have reduced skeletons and a low proportion of protein in their muscles.[5] They have probably lost the swimbladder because having a swimbladder entails replacing gas lost by diffusion. Not only will the rate of diffusion out of the swimbladder be high at the pressures at which these fish live, but the energy required to secrete a given quantity of gas will also be high. This is particularly so since the dissolved oxygen content of the sea falls to a minimum in the range of depths occupied by these fish. By losing the swimbladder, and with it the need to secrete gas, without increasing their density very much, these fish have probably saved energy. Their food supply is probably very sparse, since they live at depths where the light is dim and there is very little plankton.

Deeper still, on and near the bottom, there is both more oxygen and more food. There is more food because dead plankton which has sunk from near the surface accumulates on the bottom. It is eaten by invertebrates, which in turn are eaten by fish. These fish are much less peculiar than the deep mid-water fish. They have reasonably robust bones and muscles, and many of them have swimbladders. The macrourids are among the commonest of these benthic fish.[72] They have swimbladders, and seem to spend a lot of their time hovering or swimming just above the bottom, looking for food. Many photographs taken by cameras lowered to the depths at which they live show them just above the bottom. Some species live at depths of a few hundred metres and have retia mirabilia of normal size, but others live at 1,000–3,500 m and have very long ones.

Fish with swimbladders have been caught at 5,000 m and probably, in one case, at 7,000 m. At 7,000 m the pressure would be 700 atm, and the specific gravity of oxygen would be 0·7.[11] The swimbladder would give much less buoyancy than a swimbladder of equal volume in shallow water.

4

RESPIRATION

Selection can be expected to affect a fish's respiratory system in two ways. It will tend to fit it to the requirements of the fish, so that, for instance, a fish which evolves large red (aerobic) muscles will also evolve a respiratory system capable of providing oxygen fast enough to keep them fully active. It will also tend to keep to a minimum the amount of metabolic work which the fish must do in order to obtain a given quantity of oxygen. This chapter is concerned with both these aspects of the design of respiratory systems. It will be convenient to use teleosts to illustrate the main principles of design, but other fish will be considered briefly.

Gills

At the gills of a fish, oxygen diffuses from the water to the blood. This will only happen so long as the partial pressure of dissolved oxygen in the water exceeds the partial pressure in the blood. The rate of diffusion, R, can be expected to satisfy the general equation for diffusion.

$$R = DA. \; \Delta p/d \tag{5}$$

D is a diffusion constant whose value depends on the material through which diffusion is occurring, A is the area across which diffusion is occurring (in this case the area of the respiratory surface), Δp is the difference in partial pressures and d is the distance along which diffusion is occurring. Large values of A and small values of d will tend to give high diffusion rates. Rapid

C

uptake of oxygen at the gills depends on their having a large area, and on the diffusion distances being kept small. It also depends on the water being kept flowing over the gills. If the water were stationary, Δp would decline to a very small value. We shall see how these requirements are met.

In normal teleost respiration, water is taken in through the mouth and expelled through the two opercular openings. The gills separate the buccal cavity from the opercular cavities (fig. 9A), so the water passes through them. There are four gill arches on each side. They are curved, convex posteriorly, and each is supported by a series of bones flexibly joined together. Fig. 9B shows two arches, cut horizontally. Each arch bears a large number of plates, called gill filaments, which project laterally and posteriorly from it. Half the filaments point more laterally than the others, so that each arch bears two distinct rows of filaments. Each filament in turn bears a large number of small secondary lamellae which project up from its upper surface and down from its lower one. The gill filaments of adjacent arches touch each other at their distal ends. The secondary lamellae of one filament interdigitate to some extent with those of the next (fig. 9C). Consequently, water travelling from the buccal cavity to the opercular cavity must pass through the small channels between one secondary lamella and the next. The water in contact with the lamellae is thus continuously renewed.

The secondary lamellae consist of an epithelium, a basement membrane and an endothelium enclosing a blood space (fig. 9D).[57] The wall of the lamella is usually only $1–3\mu$ ($1–3 \times 10^{-4}$ cm) thick. The endothelium is peculiar. It consists of a series of pillar

---

Fig. 9 Diagrams showing the structure of teleost gills and the advantage of countercurrent flow. (A) Horizontal section through the head of a teleost; *am*, adductor mandibulae; *gf*, gill filaments; *o*, operculum; *p*, pectoral girdle; *ph*, pharyngeal tooth plate; *s*, suspensorium. (B) Horizontal section through two gill arches. An arrow shows how water flows between the secondary lamellae. The path of the blood is indicated on the filament at the bottom: *gr*, gill raker. (C) Vertical section through two gill filaments (*gf*) and their secondary lamellae (*la*). (D) Section through a secondary lamella: *bs*, blood space; *ep*, epithelium; *pc*, pillar cell. (E) A parallel flow heat exchanger. (F) A counter flow heat exchanger. The numbers in (E) and (F) show the temperatures of the fluids entering and leaving them. The temperatures shown in (F) are those that would be obtained by reversing the flow in the lower channel of (E) without altering the exchanger in any other way[63]

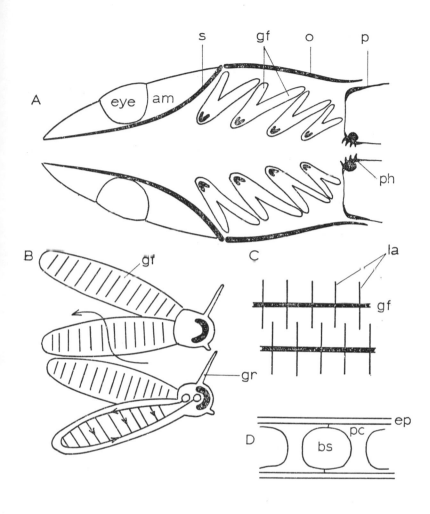

Fig. 9

cells, which hold the epithelia of the two sides of the lamella together. If the pillars were not there, the lamellae would not remain thin and flat-faced but would be inflated by the blood pressure to a bulbous shape. The pillars are reinforced by strands of basement membrane material which run through them, connecting the basement membranes of the two surfaces. The spaces between the pillar cells are just large enough to allow free passage for the red blood cells.

One edge of each gill filament adjoins the buccal cavity, and the other the opercular cavity. An afferent blood vessel runs in the opercular edge, and an efferent one in the buccal edge. These vessels are joined by the blood spaces of the secondary lamellae. Blood therefore flows from the opercular end to the buccal end of each secondary lamella. This is opposite to the direction of flow of water past the lamellae (fig. 9B). This arrangement has been compared to similar arrangements in industrial heat exchangers.[58]

Two heat exchangers are illustrated in figs 9E, F. In each a stream of hot fluid, which we will suppose enters the exchanger at 100°, loses heat to a stream of cooler fluid which enters at 0°. The pipes carrying the two fluids come into contact in the exchanger, allowing this transfer of heat. In the parallel flow exchanger (fig. 9E) the two fluids travel in the same direction. The temperature of the hot fluid falls and that of the cool fluid rises, and they approach the same intermediate temperature. If the exchanger was infinitely long, both fluids would leave it at the same temperature. The counter flow exchanger (fig. 9F) differs from the parallel flow exchanger only in having the flow reversed in one of the channels. The fluid which was initially hotter can be cooled to a temperature well below that at which the initially cooler fluid leaves the exchanger. In our example, the hot fluid is cooled to 34° and the cool fluid is warmed to 66°. If the hot fluid were to be cooled to 34° in the parallel flow exchanger, the cooler fluid would have to be made to flow more than twice as fast. For given rates of flow, more heat is transferred in a counter flow exchanger than in a parallel flow exchanger. The difference may be large or small, depending on the circumstances.[63] The arteries and veins of the swimming muscles of tunnies form a very effective counter flow heat exchanger (Chapter 2).

In a heat exchanger, heat flows from a fluid at a higher temperature to one at a lower temperature. At the gills of a fish, oxygen diffuses from a fluid where its partial pressure is higher to one where it is lower. The gills are an analogue of the heat exchanger and tend to follow mathematically similar laws (the analogy is imperfect because the partial pressure of oxygen in blood is not proportional to its concentration except over limited ranges, whereas the absolute temperature of a fluid is proportional to its heat content). The counter flow arrangement of the gills can be expected to increase the proportion of the dissolved oxygen which can be removed from the water, for given rates of flow of the water and blood.

An attempt has been made to assess the value of the counter flow arrangement of the gills, by passing water over them in the wrong direction. This greatly reduced the proportion of the dissolved oxygen removed from the water, but this is quite likely to have been due to the gills being disarranged by the reversed flow.[58]

Resting teleosts do not seem to take full advantage of the counter flow arrangement. There is evidence that some of the blood is short-circuited through channels in the gill filaments, and so does not pass through the secondary lamellae.[89] It is hard to see why the short-circuits exist. If they did not, fish could pump blood more slowly round the body and still obtain oxygen at the same rate at the gills. Perhaps some other function of the blood makes it desirable for it to be pumped faster than would be necessary for respiration alone. The short-circuits are blocked by adrenalin, so all the blood probably flows through the lamellae when the fish is active and needs to take up oxygen rapidly.

Determining the total area of the secondary lamellae is troublesome. It involves counting the gill filaments, estimating the number of lamellae on a sample of them, and measuring the areas of a carefully chosen sample of the lamellae. Nevertheless, the area has been measured for quite a number of teleosts.[55] It varies with body size and between species, but is commonly around 4 cm^2/g body weight for teleosts whose weight is of the order of 100 g. This is about twice the area of the external surface of the fish. Let us see how it is suited to meet the oxygen requirements of the fish. The difference in partial pressure of oxygen between the water and the blood will probably not be uniform all over the gills, but we can estimate the mean difference that is required for a given rate of diffusion of oxygen by re-writing equation 5 in the form

$$\Delta p = Rd/DA$$

and inserting typical values for R, d, D and A. The rate of diffusion, R, must equal the rate of consumption of oxygen by the fish. This is likely to vary between about 50 mg/Kg body weight hr in rest and 500 mg/Kg body weight hr at the maximum cruising speed (fig. 4). These rates are equivalent to 6×10^{-4} and 6×10^{-3} ml/g min, respectively. The walls of the secondary lamellae are typically about 2×10^{-4} cm thick. The gill lamellae are typically about 3×10^{-3} cm apart[55] so, if the water flows smoothly, oxygen will have to diffuse $1 \cdot 5 \times 10^{-3}$ cm or less to reach the nearest lamella. We will probably not be far wrong in taking, as an average total diffusion distance through water and tissue, 10^{-3} cm. We can ignore diffusion within the blood space because the gaps between the pillar cells are scarcely bigger than the blood corpuscles, so the plasma is probably kept thoroughly stirred by something like the 'bolus flow' that is believed to occur in blood capillaries. We can therefore take $d = 10^{-3}$ cm. The diffusion constants for oxygen in water and in tissue are different, so we will use an intermediate value, $D = 2 \cdot 5 \times 10^{-5}$/min atm.[55] The total area of the secondary lamellae is typically about 4 cm²/g body weight, but about 30% of this is taken up by the pillar cells so we will take $A = 3$ cm²/g. These values give us, for a resting fish

$$\Delta p \text{ rest} = \frac{6 \times 10^{-4} \times 10^{-3}}{2 \cdot 5 \times 10^{-5} \times 3} = 0 \cdot 008 \text{ atm}$$

and for one which is swimming at the maximum cruising speed

$$\Delta p \text{ active} = \frac{6 \times 10^{-3} \times 10^{-3}}{2 \cdot 5 \times 10^{-5} \times 3} = 0 \cdot 08 \text{ atm}$$

Δp must, of course, always be less than the partial pressure of dissolved oxygen in the water arriving at the gills. A value of $0 \cdot 08$ atm would only be possible in reasonably well oxygenated water, so our very rough calculation would lead us to suppose that the maximum rate of consumption of oxygen would only be possible in reasonably well oxygenated water. This seems to be the case. Several investigators have measured the maximum metabolic rates of active teleosts in water containing dissolved oxygen, at various partial pressures. It is low at low partial pressures, and rises as the partial pressure rises. In some species such as *Salvelinus* it goes on rising with partial pressure, right up to $0 \cdot 2$ atm (the partial pressure in water saturated with air). In others it levels off: the maximum rate of consumption of oxygen by active *Perca*, for instance, is as high at 0·1 atm as at 0·2 atm oxygen.[1]

Some teleosts such as mackerel (*Scomber*) and herring (*Clupea*) have gill areas around 10 cm^2/g body weight, which is much more than the typical 4 cm^2/g which was assumed in the calculation. These are active species. Some sluggish bottom-living teleosts such as the toadfish (*Opsanus*) have areas around 2 cm^2/g. The John Dory (*Zeus*) is a mid-water predator, but it stalks its prey instead of chasing it, and so does not need to swim fast (Chapter 5). It, too, has a very low gill area. There thus seems to be a reasonably close relationship between the gill areas of teleosts and their habits.[55]

The major part of the oxygen used by a fish which is swimming steadily at a high speed is used by the swimming muscles. The red muscles are used for sustained swimming, and are aerobic, whereas the white muscles are anaerobic (Chapter 2). The maximum rate at which a fish can use oxygen will therefore be related to the amount of red muscle which it has. One might expect the gill areas of fish with red muscle to be correlated with the weight of red muscle. Certainly mackerel and herring, which have large gill areas, have quite a lot of red muscle, but no one seems to have weighed it. The rate at which fish can recover from oxygen debt after activity of the white muscles must also depend on the gill area.

The small gill areas of sluggish fish are probably not due merely to the fish having failed to evolve bigger ones. Unnecessarily large gill areas are probably selected against, because the osmotic work which a fish has to do to maintain the ionic composition of its blood depends on the gill area. The osmotic concentration of the blood of most teleosts is about one-third of the concentration in sea water.[1] Marine fish therefore tend to lose water to their environment and to gain salts from it. Fresh water contains very low concentrations of ions, so freshwater fish tend to gain water and lose salts. Most of this diffusion between the fish and the water probably occurs at the gills, since the gill area is so large and since the blood comes so close to the water at the gills. Experiments with trout (*Salmo*) seem to show that osmotic work accounts for 20% or more of the metabolic rate, both in fresh water and in the sea,[79a] but this value may be too high.[20]

A fish which is using oxygen rapidly must pump water rapidly over its gills and blood rapidly through them. When flow is slow, the partial pressure of oxygen in the water may fall, as it passes over the lamellae, to little more than the partial pressure in the venous blood. The partial pressure in the blood may rise to quite a high value as it flows, in the opposite direction, through the lamellae. The mean difference in partial pressure between the water and the blood will be small, and the rate of uptake of oxygen will be small, but a large proportion of the oxygen will be removed from the slowly flowing water. When the water and blood flow rapidly, they will not come so near to equilibrium and the mean difference of partial pressure between them will be greater. The rate of uptake of oxygen will be greater, but the proportion of the oxygen removed from the water will be less.

This has been demonstrated by Saunders.[81] He inserted cannulae through the pectoral girdles of teleosts, into their opercular cavities, so that he could take samples of the expired water and determine its oxygen content. In some of his experiments the fish were resting, but in others he made them swim round and round in an annular aquarium. When they were resting, the fish extracted 60–85% of the dissolved oxygen from well-aerated water, but when they were active and using about five times as much oxygen they only extracted about 20–30%. To increase their rate of oxygen uptake about 5 times they must have increased the rate of flow of water over the gills about 15 times. This must be at least partly explained by the reasoning of the previous paragraph, but may be partly due to separation of the gill filaments. The gill filaments of adjacent gill arches of resting teleosts probably normally touch, as shown in fig. 9A, but they may be forced apart when the fish are active by the more rapid flow of water. A proportion of the water would then pass through the resulting gaps, and never come near the gill lamellae. It would be very difficult to observe the gill filaments of active fish, but Saunders has watched and photographed the gills of resting fish made to breathe heavily by a high concentration of carbon dioxide. When they opened their opercula, the gills could be seen with small gaps between the filaments of adjacent arches.

Respiratory pumps

Teleosts pump water over their gills by alternately expanding and contracting the buccal and opercular cavities. The mouth is open while the cavities expand and the opercula while they contract, so the water always goes in by the mouth and out by the opercular openings. Flaps of skin, just inside the mouth and along the edges of the opercula, act as valves and help to maintain one-way flow. When, for instance, the cavities start contracting, the mouth is not usually completely closed, but the buccal valves flap across its opening and prevent water from escaping through it.

To understand the mechanism of expansion and contraction of the cavities, we must study the structure of the head. Fig. 10 shows the main groups of bones which make up the skull of a typical teleost. The cranium is a rigid structure consisting of many bones firmly sutured together. It has lateral depressions which accommodate the upper halves of the eyeballs. It encloses the ears and brain. The suspensoria are more or less rigid plates, each composed of several bones. They form the lateral walls of the buccal cavity (fig. 9A) and are hinged to the cranium at the anterior and posterior ends of their dorsal edges. Between these hinge joints, they are emarginated to make room for the eyeballs. The opercula are also plates composed of several bones and have ball and socket joints with the posterior edges of the suspensoria. Each can swing through quite a large angle about the vertical axis through the joint, but the amplitudes of the other movements which the ball and socket joint permits are severely restricted by connective tissue. The lower jaws are hinged to the suspensoria and are joined to each other anteriorly by flexible connective tissue. The structure of the upper jaw (premaxilla and maxilla) varies a great deal among teleosts and will be discussed in Chapter 5 as it is important in feeding. Fig. 10 shows an arrangement which is common in primitive teleosts and is found, for instance, in trout (*Salmo*): the premaxillae are firmly attached to the cranium but the maxillae are hinged to it. The hyomandibulars are incorporated in the suspensoria but the lower parts of the hyoid arches are not. On each side there is a tiny interhyal bone and a large bar, composed of several bones more or less rigidly joined together, which we will call the hyoid bar. The hyoid bars are attached by flexible connective tissue to each other, to the opercula and to the interhyals. The interhyals are similarly attached to the median faces of the suspensoria. The branchiostegal rays of each side are attached by flexible connective tissue to the hyoid bar, and are joined to each other and to the operculum by a web of skin which forms an extension of the operculum round the curved ventral surface of the fish. The pectoral girdle forms the lateral edge of the posterior wall of the opercular cavity (Fig. 9A), and the opercular valve closes against it.

Figs. 10A-D give an indication of the range of movement of the head. In figs. 10A, C the mouth is almost closed and the buccal and opercular cavities are contracted. In figs. 10B, D the mouth is wide open (much wider than it would ever be in normal respiration) and the cavities are expanded. The suspensoria have swung laterally about their hinge joints with the cranium, carrying the opercula with them and expanding the cavities laterally. The hyoid bars have swung ventrally, lowering the floors of the cavities. Other positions are of course possible. The operculum can be opened. The mouth can be closed with the cavities expanded.

The positions of some of the muscles are indicated in figs. 10B, E. The adductor mandibulae (i) runs from the lateral face of the suspensorium to the lower jaw. When it contracts, it closes the mouth. A strip of muscle (ii), which is sometimes but not always divided into several discrete muscles (iia) (iib), runs from the cranium to the medial faces of the suspensorium and operculum. It serves to contract the buccal and opercular cavities and to close the operculum. Another muscle (iii) runs from the cranium to the lateral face of the suspensorium, and serves to expand the cavities. A muscle from the cranium to the lateral face of the operculum (iv) opens the operculum by swinging it laterally about its joint with the suspensorium. Two other important muscles are shown in fig. 10D. Muscle (v) pulls the anterior ends of the hyoid bars ventrally and posteriorly, lowering the floors of the buccal and opercular cavities. It also helps to swing the suspensoria and opercula laterally, because the posterior ends of the hyoid bars are forced laterally as the anterior ends are pulled posteriorly. Muscle (vi) has the reverse action. The mechanisms for opening the mouth are rather indirect. A ligament, shown in fig. 10B, attaches the ventral corner of the operculum to the lower jaw. If the ventral corner of the operculum moves posteriorly, the mouth is pulled open. Contraction of the posterior part of muscle (ii) (labelled iib in fig. 10B) has two effects. It tends not only to pull the operculum medially, but also to rotate it about a transverse axis through its joint with the suspensorium. On the left side of the head (which is shown in fig. 10A, B) it tends to rotate the operculum anti-clockwise. It thus moves the ventral corner of the operculum posteriorly and opens

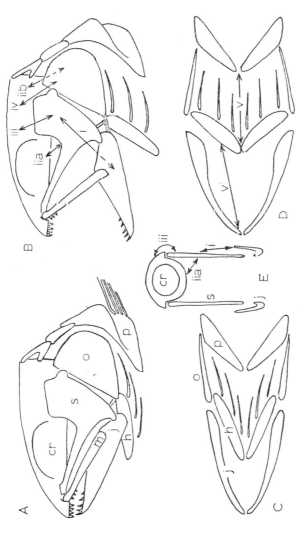

Fig. 10 Diagrams of the skull of a typical teleost. (A) Lateral view with the mouth closed and the buccal and opercular cavities contracted. (B) Lateral view with the mouth open and the cavities expanded. (C), (D) Ventral views in the same positions as (A), (B). (E) Transverse section. The two-headed arrows represent muscles, and the lines between the operculum and the lower jaw in (B) represent a ligament. *cr*, cranium; *h*, hyoid bar; *j*, lower jaw; *m*, maxilla; *o* operculum; *p*, pectoral girdle; *s*, suspensorium

the mouth. When muscle (v) contracts the hyoid bar pushes posteriorly as well as laterally on the operculum and so tends to open the mouth. The mouth opens whenever either muscle (iib) or muscle (v) contracts while muscle (i) is relaxed.

Ballintijn & Hughes[18] investigated the use of the muscles in respiration. They fastened lightly· anaesthetised trout (*Salmo*) in a holder, and stuck electrodes into various muscles. They were able to record the muscle action potentials and so to find out which muscles were active at each stage in the respiratory cycle. Gentle respiration seems to depend entirely on muscles (i), (ii) and (iii). Muscle (i) contracts, closing the mouth. While it is still contracting, muscle (ii) contracts, reducing the volumes of the buccal and opercular cavities and forcing water out of the opercula in spite of the tension in (iib). Eventually (i) relaxes, and (ii) is able to perform its other function of opening the mouth. When the mouth is open, (ii) relaxes and (iii) contracts, enlarging the buccal and opercular cavities and drawing water into the mouth. Although the opercula are carried laterally with the suspensoria, the flexible valves on their edges remain closed against the pectoral girdle and prevent water from entering. Muscle (i) then contracts and the cycle is repeated.

Muscles (iv), (v) and (vi) are used, as well as the others, in deep breathing. (v) helps first to open the mouth and then to expand the cavities. (vi) helps to contract the cavities. One might suppose that (v) and (vi) would contract simultaneously to open the mouth, but they seem always to be used at different stages in the cycle, and never together.

When a fish is swimming, it can pass water over its gills simply by keeping its mouth open. The mackerel, *Scomber*, relies on this to such an extent that the oxygen content of its arterial blood falls to a very low value when it is kept in well-aerated water but prevented from swimming.[1] It makes no breathing movements as it swims, but at least some other teleosts make strong breathing movements when they are swimming.[81]

In respiration, water is pumped through the fine channels between the secondary lamellae of the gills. This requires energy, and the energy will be least if the rate of flow is kept constant. The pressure required to drive a fluid through a narrow tube is proportional to the rate of flow.[91] The work done in driving the fluid through the tube is the product of the volume that passes through and the difference in pressure between the two ends (this is a

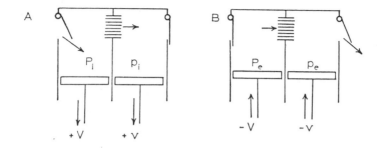

Fig. 11   The model of the respiratory pump described in the text showing (A) inspiration and (B) expiration. These diagrams are based on diagrams by Hughes[58]

corollary of the definition of work as the product of the force and the distance moved along the line of action of the force). When the rate of flow increases, so does the work required to drive a given volume of fluid through. It therefore requires more work to drive a quantity of fluid through the tube in a series of short spurts than to drive it through at a steady rate over the same period.

We will consider the work required for respiration more closely. It will be convenient to refer to a simplified model of the respiratory pump (Fig. 11). The chamber on the left of each diagram represents the buccal cavity, and the chamber on the right the two opercular cavities. They are connected by a battery of fine tubes which represents the gills. The volume of the buccal cavity can be increased and decreased through a range V, and the combined volumes of the opercular cavities through a range v. This is represented as being done by the action of pistons. The differences in pressure between the cavities and the water surrounding the fish are P_i, p_i in inspiration, when the cavities are increasing in volume (fig. 11A), and P_e, p_e in expiration, when the cavities are decreasing in volume (fig. 11B).

In inspiration, the volume of the buccal cavity increases by V, and the volume of the opercular cavities by v. A volume ($V + v$) of water enters the mouth, and the pressure in the buccal cavity falls below the pressure in the surrounding water, because work must be done to draw the water into the mouth. This work is required mainly for accelerating the water. Its amount is

$- P_i(V + v)$ (this is a positive amount of work since P_i is negàtive).
Of the volume $(V + v)$ which enters the mouth, v is immediately
drawn through the gills to the opercular cavities. The pressure in
the opercular cavities falls below the pressure in the buccal
cavity, and work $(P_i - p_i)v$ is done in drawing this water through
the gills. This work is done mainly against viscosity, since the
channels between the secondary lamellae are so fine. In expiration, the volumes of the buccal and opercular cavities decrease by
V and v, respectively. The pressures in the cavities rise above the
pressure in the surrounding water. A volume of water, V, from
the buccal cavity is driven through the gills, and work $(P_e - p_e)V$
is done in the process. The whole volume, $(V + v)$, is driven out
of the opercular openings and work $p_e(V + v)$ is done, mainly in
accelerating it.

A volume v of water passes through the gills in inspiration,
and a volume V in expiration. Let inspiration take time T_i and
expiration T_e. Then, if the rate of flow of water over the gills
is to be constant

$$v/T_i = V/T_e \qquad\qquad (6)$$

When the rate of flow is constant, the pressure difference across
the gills will also be constant, and

$$(P_i - p_i) = (P_e - p_e) \qquad\qquad (7)$$

It is therefore possible to find out whether the rate of flow is
constant by measuring the pressures in the buccal and opercular
cavities. It is of course necessary to use instruments which can
follow rapid fluctuations of pressure. Ordinary manometers would
not be much use.

Such measurements have been made. Hughes and Shelton[58]
fastened lightly anaesthetised fish in a holder, and slipped tubes
from their pressure measuring devices through the mouth and
one of the opercular openings. Saunders[80] passed his tubes
through the skull and pectoral girdle, and so indirectly into the
buccal and opercular cavities. The pressure records are not as
simple as our model (fig. 11) would suggest. The pressures in the
cavities do not change instantaneously from steady values P_i, p_i
to steady values P_e, p_e as inspiration ends and expiration begins,
or vice versa. Instead, the pressure in each cavity oscillates
fairly smoothly between a minimum value attained in inspiration
and a maximum value attained in expiration. Further, the buccal

and opercular cavities do not enlarge and contract absolutely simultaneously. They are usually out of phase by about 0·1–0·2 cycles, with the buccal cavity leading. There is usually a short phase when the buccal cavity starts expanding while the opercular cavities are still contracting and the pressure difference across the gills is reversed. It probably occurs because the mouth must be opened and the opercula closed at the beginning of inspiration. Opening the mouth will tend to increase the volume of the buccal cavity and closing the opercula will tend to reduce the volumes of the opercular cavities. Probably very little back flow results, as the reversed pressure difference is small, but the forward flow of water through the gills must be interrupted. This interruption must increase the work required for respiration, since the work is least when the rate of flow is constant. Apart from this interruption, many species maintain a reasonably constant pressure difference across the gills, and so presumably a reasonably constant rate of flow. They include some species such as the roach, $Rutilus$, which expand their cavities quickly and contract them slowly ($T_i < T_e$ in equation 6 and so, presumably $v < V$). They also include species such as the whiting ($Gadus\ merlangus$) which expand the cavities slowly and contract them quickly ($T_i > T_e$ and, presumably, $v > V$).

The measurements show that pressure differences of around 0·3 cm·water are usually enough to suck water into the mouth and to drive it out through the opercula of resting teleosts, while pressures of 0·5–1 cm water are needed to drive it through the gills. $-P_i$ and p_e are usually small compared to ($P_i - p_i$) and ($P_e - p_e$). They can of course be kept small by opening the mouth and opercula reasonably widely. Correspondingly small amounts of work are involved. It can, however, be advantageous to do a considerable amount of work in accelerating the water out of the opercular openings. Some fish rest on the bottom in still water for long periods. When they do this, their respiration must tend to reduce the oxygen content of the water immediately surrounding them, though no one seems to have measured this effect. The effect can be made less serious by expelling the expired water in a fast jet which will travel some distance from the fish. Sluggish bottom-living fish tend to have opercular cavities which expand and contract much more than the buccal cavity (i.e. $v > V$) so that they can expand the opercular cavities slowly and contract them rapidly and yet maintain a reasonably constant rate of flow through the gills. The opercular openings are usually small, so that the water must pass through them at high velocity. They are usually dorsal in position, so that the water travels upwards and does not make the fish conspicuous by disturbing the sand around it. The dragonet, *Callionymus*, shows these adaptations. It takes about four times as long to expand the cavities as to contract them, and pressure records indicate that nearly as much work is done in accelerating the water out of the small opercular openings as in pumping it through the gills.

We will now estimate the amount of energy used by fish in respiration, so that we may have some idea of its importance in the total energy requirements of fish. We will calculate the amount of oxygen required for the metabolism of the respiratory muscles, as a proportion of the total amount of oxygen obtained by respiration. We will consider first a fish resting in air-saturated water.

We can calculate the amount of work done by the respiratory muscles from the available records of pressures in the buccal and opercular cavities.[58, 80] If the rate of flow through the gills is constant so that equation (7) is satisfied, we can write, for our simple model

$$(P_i - p_i) = (P_e - p_e) = (P - p)$$

and the total work done in the model in each respiratory cycle must be $(-P_i + P - p + p_e)(V + v)$. The work W, which must be done in breathing a unit volume of water, is therefore given by the equation

$$W = -P_i + P - p + p_e \qquad (8)$$

Fish are not as simple as the model, as we have seen, but a reasonable estimate of W can be made from the records if account is taken of the fact that the water will flow fastest when the pressure differences are highest. It appears that, for most of the species that have been investigated breathing quietly in well-aerated water, the total of pressures (equation 8) is between 1 and 1·5 cm water, so W is between 1,000 and 1,500 ergs/cc (a pressure of 1 cm water equals 1,000 dynes/cm^2, and an erg is a dyne-cm). We will take an intermediate value and so estimate that it takes $1,300 \times 1,000 = 1\cdot3 \times 10^6$ ergs to breathe 1 litre of water. This is equivalent to 0·03 cal, but the muscles are unlikely to be more than 20% efficient so the metabolism that occurs in them is likely to be about 0·15 cal for every litre of water. 1 ml oxygen oxidises food to yield about 5 cal, whatever the food, so the respiratory muscles will need about 0·03 ml oxygen for every litre of water that is passed over the gills. A resting fish in air-saturated fresh water obtains about 6 ml oxygen from every litre of water,[81] so a mere 0·5% or so of the oxygen which is taken up is required by the respiratory muscles.*

We will now consider an active fish. We have seen that a swimming fish using oxygen 5 times as fast as a resting one must pass water 15 times as rapidly over its gills. Hughes and Shelton[58] have measured the pressure required to drive water at various speeds through the gills of a tench (*Tinca*), which was so heavily anaesthetised that the respiratory movements had stopped. Their results indicate that an increase of 15 times in the rate of flow would require an increase of at least 10 times in the pressure difference across the gills, and so an increase of at least 150 times in the power required to drive the water through the gills. No measurements seem to have been made of the pressures in the buccal and opercular cavities of swimming fish, but it seems likely that the power used in respiration must increase about 150 times to increase the rate of uptake of oxygen 5 times. Respiration will then require not 0·5% but 15% of the oxygen uptake.

So far we have only considered respiration in well-aerated water, containing about 8 ml dissolved oxygen/l. Now we will consider respiration in water containing only 1 ml oxygen/l. The resting metabolic rate could only be maintained at the same level as in aerated water (where 6 ml can be extracted from every litre) by increasing the rate of flow about 20 times, since only about 0·3 ml oxygen would probably be obtained from each litre of water.[81] Such an increase in the rate of flow would probably require an increase of more than 200 times in the work required for respiration. Our fish, striving to keep its rate of oxygen uptake at the level found in well-aerated water, would require not 0·5% but all of the oxygen it obtained to do the work of respiration. If the fish allowed its total oxygen consumption to

---

* A much higher estimate has been obtained by comparing the metabolic rates of lightly anaesthetised fish breathing naturally with those of fish whose gills were irrigated artificially after the respiratory muscles had been immobilised with succinylcholine.[83] It is suspect because succinylcholine blocks nerve-muscle junctions and probably reduced tonus throughout the bodies of the fish.

drop by 50%, its respiratory muscles would require much less energy and some of the oxygen obtained would probably be available for non-respiratory purposes. There is probably an optimum level of respiratory effort in poorly oxygenated water beyond which further increases in effort leave less, rather than more, oxygen available for non-respiratory purposes. If this optimum effort does not supply enough oxygen to keep the fish alive, it must die. A selection of British freshwater species die at oxygen concentrations below 0·3 to 1·5 ml/l.[37]

There is more direct evidence that respiration accounts for a large proportion of the total metabolism at low oxygen concentrations. Beamish found that the resting metabolic rate of various freshwater fish remained more or less constant as the oxygen content of the water was reduced from air saturation to about 5 ml/1.[21] As the oxygen content was reduced further, the metabolic rate rose by about 50% before finally declining at the lowest oxygen contents. Presumably the rise was due to increased respiratory effort.

It thus appears that respiration requires quite a large proportion of the total metabolism of swimming fish, and a very large proportion of the total metabolism of fish in water with a low oxygen content. Characters which affect the metabolic cost of respiration may be important in determining the maximum sustained swimming speeds of fish, and their ability to survive in low concentrations of oxygen. The metabolic cost of gentle respiration, such as suffices for fish resting in well-aerated water, is probably very low. Nevertheless, a genotype whose only phenotypic effect was to reduce it might be sufficiently favoured by natural selection to spread reasonably rapidly through a population. This point was discussed quantitatively in Chapter 1.

Work is of course done in driving the blood through the secondary lamellae as well as in driving the water between them. A resting cod (Gadus) pumped through its heart about 600 ml blood/Kg body wgt hr.[59] The pressure of the blood of such teleosts as have been investigated falls by about 2cm Hg or 3×10^4 dynes/cm² as it passes through the gills.[1] We may therefore estimate that the work a resting teleost does driving blood through its gills is likely to be about $600 \times 3 \times 10^4 = 1.8 \times 10^7$ ergs/Kg body wgt hr. If heart muscle is 20% efficient, this would require metabolism involving about 0·6 mg oxygen/Kg body wgt hr. This would probably be 1–2% of the resting metabolic rate.

In a few teleosts, the respired water follows unusual paths. *Gyrinocheilus* (Cyprinoidei) is one of the fish with suckers referred to in Chapter 3. Its mouth is in the middle of the sucker and is therefore blocked when the sucker is in use. Water is breathed in through a dorsal part of the opercular opening, whose valve is reversed so that it can only admit water, and out through the more ventral part of the opening. This can be·demonstrated by releasing a suspension of carmine particles near a resting specimen.

The flatfish (Pleuronectiformes) lie with one side on the bottom and the other (which bears both eyes) facing upwards. They often bury themselves, apart from the mouth, eyes and upper operculum, in sand. If water was expelled from the lower operculum it would have to percolate through the sand. Considerable pressure (p_e) would probably be required, and the sand might be disturbed. The flatfish have evolved a channel, absent in other fish, between the two opercular cavities. A flatfish resting quietly in an aquarium with a transparent bottom can be seen to be moving both opercula, but if a suspension of carmine particles is released near the mouth, any that enters the mouth is expired from the upper operculum and none from the lower one. It seems that flatfish resting on the bottom pump water over both sets of gills, but the water from the lower opercular cavity travels through the channel to the upper one, and all the water is expired from the upper opercular opening.[94]

Other fish

So far we have considered only teleosts, though much of what has been said applies equally to other fish, and some features of the respiratory system are shared by other classes of fish. Lampreys, hagfishes and selachians, like teleosts, have a series of openings from the pharynx to the exterior, with gill filaments and secondary lamellae. The blood flows through the lamellae in the direction opposite to the flow of water over them. In other respects they are unlike teleosts. The respiratory apparatus and mechanisms of lampreys and hagfishes are so different from each other and from those of other fish that they cannot be described briefly and will not be described here. A good account of them is given by Hughes and Shelton.[58] The selachians, however, require description. The most striking difference between their respiratory system and that of teleosts is that selachians have no operculum. Instead, they have a flap of muscle and skin, reinforced by cartilage, extending laterally from each gill arch. This flap extends beyond the filaments and turns posteriorly to form, as it were, an individual operculum for the gill slit immediately posterior to its arch. Whereas a teleost has one opercular cavity on each side, a typical selachian has five parabranchial cavities (a few have more than five gill slits and correspondingly more cavities). Water is driven over the gills by expansion and contraction of the buccal and parabranchial cavities.

The mechanisms of respiration of the dogfish (*Scyliorhinus*) has been investigated by recording the electrical activity of its muscles.[56] There is a group of muscles, the hypobranchial muscles, which can be used to open the mouth and expand the buccal and parabranchial cavities, but they are not normally used in respiration. Certain tissues of the head are elastic, and their elasticity tends to open the mouth and expand the cavities. The only muscular activity in gentle respiration seems to be contraction of, first, the adductores mandibulae, which close the mouth, and then of the many muscles which constrict the buccal and opercular cavities. The mouth opens, and the cavities expand, elastically. A little activity can be recorded from the adductores mandibulae long before the mouth starts closing. Apparently these muscles are used to check the elastic opening of the mouth as well as to close it.

The reliance on elasticity seems to have a disadvantage. The work of respiration will be least if the water flows over the gills at a constant rate. A constant rate of flow requires, among other things, a constant rate of expansion of the parabranchial cavities. However, if the cavities expand elastically, they can be expected to expand relatively rapidly at first, and then more and more slowly as they approach the equilibrium volume and the elastic restoring force declines. This seems to be what happens. The pressure difference across the gills reaches a peak early in inspiration and then dies gradually away.

Some sharks stop their respiratory movements when they swim, and maintain a current of water over the gills simply by keeping their mouths open.

Selachians have a rudimentary gill opening, the spiracle, in front of the hyoid arch. Some of the water passes through it in inspiration, and a valve closes it in expiration. Brachiopterygii and sturgeons have spiracles, but teleosts do not.

The rays have ventral mouths and gill openings, but large dorsal spiracles. Many of them, like flatfish, rest on the bottom partly buried in sand. On the bottom, they breathe in entirely through the spiracle, and so avoid taking in sand with the water. They breathe out through the ventral gill openings. They do not seem to have any adaptation comparable to the channel between the opercular cavities of flatfish, to avoid the necessity of breathing out into the sand.

Air breathing

Ordinary teleosts asphyxiate in water containing less than about 0·3–1·5 ml oxygen/l.[37] Activity is limited even at much higher oxygen concentrations; for instance, the cruising speed of *Perca* declines quite sharply below about 2·5 ml/l.[1] Low oxygen concentrations occur in various conditions in nature. Concentrations below 1 ml/l are found in the deeper waters of lakes in summer, at the oxygen minimum (well below the surface) in some parts of the oceans, and in some ponds and swamps. Tropical swamps may have very low oxygen concentrations even within a centimetre of the surface, but the water at the surface must always have a reasonable oxygen content because it is in contact with the atmosphere. Some very small teleosts survive in stagnant tropical swamps by swimming just below the surface, breathing this water.[1]

Fish in ponds and swamps are near enough to the surface to visit it and breathe air. Many have evolved the ability to do so. Gills do not normally form satisfactory air-breathing organs. When they are immersed in water the filaments and lamellae are supported by the water and remain separate, but when they are taken out of water the filaments fall together and are held together by the surface tension of the moisture on them. A gill in air therefore presents a relatively small surface area to the air, and the surfaces of most of the lamellae are separated from the air by long diffusion paths through stagnant water. Nevertheless a few teleosts with apparently normal gills use them for breathing air and can survive in water which contains very little dissolved oxygen. When they are kept in such water they visit the surface every few minutes to re-fill the opercular cavities with air. The visits to the surface of these and other air-breathing fish are much less frequent than the pumping movements of aquatic respiration, largely because of the high oxygen content of air. A litre of air contains about

200 ml oxygen but a litre of water saturated with air at 10°C contains only 8 ml dissolved oxygen. A mouthful of air thus contains 25 times as much oxygen as an equal volume of air-saturated water. Further, air-breathing fish seem to take in a volume of air at each visit to the surface which is larger than the volume of water passed over the gills in each cycle of gentle aquatic ventilation. *Hypopomus,* which lives in swamps in tropical South America, is one of the few fish known to use normal gills for breathing air.[1]

The catfish *Clarias* lives both in swamps and in better-oxygenated waters, in Africa and South Asia. The ventral parts of its gills are normal but the dorsal parts are remarkably modified for breathing air. The secondary lamellae of the dorsal parts are squat and they are borne on structures which, unlike ordinary gill filaments, do not collapse in air. Groups of filaments are fused together to form fairly stiff fan-like structures. The second and fourth gill arches bear large arborescent structures which are supported by cartilage. The fan-like structures, the arborescent ones and the dorsal wall of the opercular cavity bear the squat lamellae.[2,10] The Asiatic labyrinth fishes (*Anabas,* etc.) have rather similarly modified gills, and breathe air.[6]

We saw in Chapter 3 that the swimbladder probably first evolved as a lung, and that it retains this function in the dipnoans, holosteans and Brachiopterygii. These lungs have a rich blood supply from the last aortic arch or (in *Lepisosteus*) from the coeliac artery.[2] They have a dense network of capillaries which is only separated by a thin epithelium from the air in them. Their internal surface area is greatly increased by a network of ridges which is sometimes so complicated as to give the inside of the lung a sponge-like appearance. Since the ridges form a network, they do not fall together like gill filaments in air. A number of teleosts use their swimbladders as lungs and it seems probable that in at least some cases the respiratory function has been regained after being lost. An example is *Erythrinus,* from forest swamps in tropical South America. Its swimbladder is in many respects just like those of other Cypriniformes. It has a swimbladder duct (which is of course essential to its functioning as a lung) and is divided into two chambers of which the more anterior is connected to the ears by Weberian ossicles (Chapter 6). However, the posterior chamber is lung-like, and is used as a lung.

The gas in the lung of *Lepisosteus* has been analysed. Samples taken immediately after air had been taken at the surface contained on average 7·2% oxygen, and samples taken shortly before the next visit to the surface was due 3·8%. The amount of oxygen obtained by the fish from each breath must be at least 3·4% of the volume of the lung, or about 2·8 ml oxygen/Kg body weight. This seems to be about 8% of the hourly resting metabolic rate. The low proportion of oxygen in the lung immediately after taking a breath implies that no more than about a third of the contents of the lung is changed at each breath.[11]

Some teleosts such as the mudskipper, *Periophthalmus*, have a rich blood supply to the wall of the buccal cavity, and use it for breathing air. Others, such as the catfish *Hoplosternum* (which lives in the same swamps as *Erythrinus*), have part of the gut modified to serve as a lung.

Some air-breathing fish only breathe air when the concentration of dissolved oxygen in the water is low. *Erythrinus* is an example.[1] The oxygen concentration at which it starts breathing air depends on the carbon dioxide concentration, but is never less than about 1·5 ml/l. There is a range of oxygen and carbon dioxide concentrations over which aerial and aquatic respiration are used together, but when there is less than about 0·5 ml oxygen/l or when the carbon concentration is very high or (surprisingly) very low, only aerial respiration is used. The oxygenated blood from the swimbladder is returned to the heart, and travels through the gills before going to the rest of the body. If water of very low oxygen content was pumped over the gills, much of the oxygen taken up at the swimbladder would be lost again from the gills. The partial pressure of oxygen in water containing 0·5 ml/l at the temperature at which *Erythrinus* lives is about 0·02 atm. This is well below the partial pressure at which typical teleost blood becomes saturated (fig. 8B). The properties of *Erythrinus* blood have not been investigated.

Carbon dioxide is about 30 times as soluble in water as oxygen. For this reason, the partial pressure of carbon dioxide in water passing over the gills of a fish increases far less than the partial pressure of oxygen decreases, in aquatic respiration. The blood of normal teleosts is not exposed to high partial pressures of carbon dioxide at the gills. However, the stagnant waters in which air-breathing fish live often contain high partial pressures of carbon dioxide. (Partial pressures around 0·03 atm occur in South American swamps.) Carbon dioxide can only be lost from the gills after its partial pressure in the blood has built up to a higher value. Air-breathing fish lose some of their carbon dioxide by diffusion to the air in the respiratory organ,[1] but this process leads rapidly to high partial pressures in the organ. If they lost all their carbon dioxide in this way, the partial pressure of carbon dioxide in the organ would rise almost as much as the partial pressure of oxygen fell, between one breath and the next (it would not rise as much because respiratory quotients are usually less than 1). Whether a fish living in water with a high carbon dioxide content loses its carbon dioxide to the water or to the air or to both, it must have a high partial pressure of carbon dioxide in its blood. In normal fish, this would reduce considerably the amount of oxygen which the blood could take up, owing to the Bohr and Root effects (fig. 8B). These effects are exceptionally slight in fish from tropical swamps, whether they breathe air or not.[1]

Fish that breathe air commonly have small gill areas. We have seen that an unnecessarily large gill area is disadvantageous, because of the relationship between gill area and osmotic work.

Air-breathing not only enables fish to survive in water containing little oxygen; it also enables them to leave the water. *Clarias*, for instance, sometimes makes journeys overland at night. *Periophthalmus* spends much of its time out of water on mud flats. The African lungfish *Protopterus* often lives in swamps which dry up in the dry season. It survives, breathing air, in a burrow which it digs in mud.[6]

5

FEEDING

The relationship between food, growth and fecundity was discussed in Chapter 1. An inherited character which enables its possessors to get more food will generally enable them to grow faster and to produce more offspring before they die. It will generally be favoured by selection. This chapter is about the ways in which fish obtain their food and swallow it, and about the structures and mechanisms they use in these processes. Most of the chapter is about bony fish, but other fish are discussed at the end.

Hunting methods

Before a fish can take food into its mouth it must get very close to the food. This presents special problems when the prey is active. Even if the predator can catch its prey by simple pursuit, this is liable to use up a lot of energy. In so far as hunting techniques are inherited, they must tend to evolve so as to become more effective, and so as to use less energy.

Many predatory fish ambush their prey or approach it stealthily, rather than pursue it openly. The pike, *Esox*, lives among underwater plants and approaches fish that come near cautiously, by means of fin movements, before making a short final dash at them. Various groupers (*Epinephalus*, etc.) lie on the bottom of the sea, sometimes concealed by coral or rocks, until an unwary fish or crustacean comes near enough to be captured by a short dash.[53] *Synodus* buries itself in sand with only its eyes and the top of its head exposed, and darts out to catch prey that comes near.[53] The John Dory, *Zeus*, approaches its prey slowly by fin movements. Its body is so compressed that it is inconspicuous when seen head-on. When it is near enough it opens its mouth suddenly, and sucks the prey in.[6]

The angler fish *Lophius* is well known for its remarkable method of catching fish.[3] The first ray of the dorsal fin (a spiny ray) is very long and has large muscles. Its base lies well forward on the top of the angler's head and its distal end bears a tag of skin (the 'bait') which is more conspicuous than the ray itself. The angler lies in a hollow on the bottom, very well concealed by its excellent camouflage. When potential prey approaches, the angler moves the bait, which the prey may investigate as possible food. If it does, the angler keeps the bait moving, just ahead of the prey, and finally swings it down anteriorly. The prey, if it has followed the bait, is then swimming head down just in front of the angler's mouth. The angler springs forward and seizes it. A group of angler fish related to *Lophius* swim in mid-water, at depths of 1,000 m and more below the surface of the sea, where it is dark even by day. They have luminous baits, which they presumably use in much the same way as *Lophius* does.[5]

Mouths of bony fish

Fish eat some foods complete. When they eat insects and other fish, for instance, they usually take the whole of the prey as a single mouthful. Bites are taken from other foods, such as leaves of plants. In either case the food must be taken into the mouth. There are three ways in which this can be done.[13]

(1) The fish may swim up to the food with its mouth open, so that the mouth comes to enclose the food. The food is not sucked towards the mouth as in methods (2) and (3), but the mouth moves round it. Films of *Euthynnus* (Scombridae) taking pieces of fish in mid-water show that it feeds in this way.[93] *Lepisosteus* (fig. 2) has very long jaws and spiky teeth, and feeds on smaller fish. It swims to one side of its prey, opens its mouth and flicks its head sideways to get its jaws round the prey, and grasps the prey between its teeth. It swallows the prey by relaxing its grip and swimming forwards. The anchovy (*Engraulis*) does not suck food into its mouth when it is feeding in a dense patch of plankton. It merely swims forward with its mouth and opercula open. The

water flows in through the mouth and out through the opercular openings, and the plankton is filtered out of it by the gill rakers which will be described later.

(2) A fish may suck food into its mouth by enlarging its buccal and opercular cavities. The fish is more or less stationary as the food enters its mouth. This seems to be the normal method of taking food from the bottom. The angelfish (*Pterophyllum*) shown in fig. 12 are feeding in this way. In fig. 12A the angelfish is approaching a piece of earthworm on the bottom of its aquarium, with its mouth closed and its buccal and opercular cavities contracted. In fig. 12B it has opened its mouth to take the food, and the cavities are expanding to suck it in. Their expansion is indicated by a bulge in the ventral profile which is due to depression of the hyoid bars, and by the spreading of the branchiostegal membrane (compare fig. 10B). In fig. 12C the food is in the mouth, which has been closed. The buccal and opercular cavities are still expanded. Obviously they must not contract before the mouth closes, or the food would be blown out again.

(3) Some fish get food into their mouths by using methods (1) and (2) together. The fish swims towards the food and sucks at the same time. Pike seem to do this, and many other predators which feed in mid-water probably do the same.

In vigorous respiration the mouth is opened by the muscles labelled (iib) and (v) in fig. 10, and the buccal and opercular cavities are expanded by muscles (iii), (iv) and (v). Experiments with perch (*Perca*) fitted with electrodes have shown that the same muscles are used in feeding.[77b] At the same time the dorsal swimming muscles contract, raising the cranium and upper jaw while the lower jaw is lowered. Feeding involves much larger and briefer pressure changes than respiration. Teleosts have been trained to suck rings of food off the end of a fine tube connected to an instrument which can record rapid changes of pressure. The movements they used were very like those of normal feeding. They had to put their mouths round the end of the tube as they sucked the food off, so the pressure in the buccal cavity was recorded. This pressure was reduced by 50–100 cm water for about $\frac{1}{20}$ sec as orfe (*Idus*) sucked food in.[14b] Most species investigated so far give similar records but some are able to suck even harder.

In respiration, it is desirable that all the water that goes into the opercular cavities should travel through the fine channels between the secondary lamellae of the gills. Most of the work done by the respiratory pump is used in driving the water through these fine channels. In feeding, the gills are a hindrance. If their fine channels could be by-passed, muscles of limited power could expand the opercular cavities faster and suck the prey faster into the mouth. The faster it is sucked into the mouth, the less chance active prey has of escaping. Normally, the gill filaments of successive arches touch (fig. 9A) but there are muscles which can separate them.[58] It seems likely that when a fish sucks food into its mouth, it may separate the gill filaments so that there are large openings through which water can pass from the buccal cavity to the opercular cavities.

The size of a fish's mouth limits the size of pieces of food which can be eaten whole. Many predatory fish eat fish which are up to 40% of their own length or about 6% of their volume, and need rather large mouths.[1a] Some marine fish which swim in mid-water at great depths can eat prey of the same volume as themselves.[6] They can open their mouths exceedingly widely and have hugely distensible bellies.

Where a large mouth is not necessary, a small one may be advantageous. Fig. 12D shows how water flows when it is sucked into a tube. It converges on the mouth of the tube from all directions, its speed increasing as it does so. Water must flow similarly into a fish's mouth when the fish sucks food in. It will attain its maximum speed as it passes through the mouth opening. If the gill filaments are separated as we have supposed, most of the work done by the muscles is probably used in accelerating the water to this speed. The power involved is the amount of kinetic energy that is given to the water in unit time. If the water enters a mouth opening of area A sq cm with a velocity u cm/sec, the power required is $\frac{1}{2}Au^3$ ergs/sec. If the power available is constant, a smaller mouth will allow a greater speed to be attained. The sea horse (*Hippocampus*, fig. 1) feeds on small planktonic crustaceans. To catch them, it must suck them into its mouth faster than they can swim away. It has a very small mouth which is big enough to admit the crustaceans and which presumably enables it to suck them in rather fast.

When a fish takes inactive food, the speed at which it enters the mouth is unimportant, but the distance from which it can be sucked may be important. It will be important if access to the food is limited in any way—if, for instance, the food is an insect larva resting in the cleft between two pieces of gravel. It will

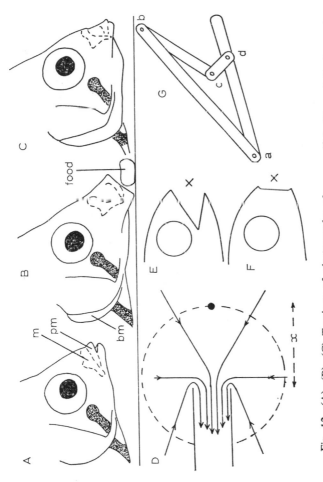

Fig. 12 (A), (B), (C) Tracings of photographs of an angelfish (*Pterophyllum*) feeding[13]. The positions of the premaxilla (*pm*) and maxilla (*m*) are indicated by broken outlines. *bm*, branchiostegal membrane. (D), (E), (F), (G) Diagrams which are explained in the text

obviously depend on the volume of water that can be sucked into the mouth. Consider fig. 12D again. If the particle shown at a distance x from the mouth of the tube is to be sucked into the tube, all the water enclosed by the broken line must be sucked in as well.

The distance from which a fish can suck food into its mouth can be estimated if the volume of water it can suck in is known and if certain assumptions are made.[13] One must assume that the food has the same density as the water and moves at the same speed as the surrounding water. This is obviously not true when the food is a thick-shelled mollusc which must be sucked up faster than it can sink, but is a reasonable assumption in most cases. It is convenient to assume that water flows towards the mouth equally from all directions as shown in fig. 12D, which is not quite true when the fish is feeding from the bottom. One must assume that the flow of water stops as soon as the buccal and opercular cavities have reached their maximum volumes, although it is possible that the opercula might open and allow the flow to continue for a moment by inertia. None of these assumptions seem likely to cause very large errors.

It has been estimated from measurements made on cinematograph films of orfe (*Idus*)[14b] and some other species, that the volume of water sucked in as the buccal and opercular cavities enlarge in feeding is about 6% of the volume of the body. This is similar to the maximum expansion in respiration. It seems to be sufficient to suck in food from distances of up to about one-quarter of the length of the head. This estimate depends on the assumptions listed above. Observations and photographs of fish feeding suggest that it is about right for typical teleosts.

Teleosts probably vary in the amounts by which they can expand their buccal and opercular cavities. There are certainly differences in the sizes of their mouths. When the mouth opening is very big the maximum distance from which prey can be sucked will be small, but as it decreases in size there must come a point when further decreases make negligible differences to the distance. This is because water flows towards the mouth from all directions (fig. 12D). If the diameter of the mouth opening is smaller than the distance x of the food from the mouth, the volume of water that must be sucked into the mouth is roughly the volume of a sphere of radius x and is nearly independent of the size of the mouth.

Figs. 12E, F represent two fish with differently shaped mouth

openings. The fish in fig. 12E has a mouth that opens in something
like a grin. The mouth of a pike opens like this. The fish in fig.
12F has a round mouth opening like that of the angelfish in
fig. 12B. The round type of mouth seems better adapted than
the grinning type for sucking in food. Suppose that the fish in
figs. 12E, F suck particles of food into their mouths from the
positions X, which are equidistant from their snouts. The fish
with the grinning mouth will have to suck in a greater volume
of water than the fish with the round one, because water will
enter at the corners of its mouth. However, the grinning type
of mouth can have an advantage in catching fish. A fish with the
round type of mouth cannot grip its prey sideways on. Perch,
which have the round type of mouth, have to manœuvre to
get their prey head on before capturing it. Pike can seize their
prey at any angle and then work it round to the right position
for swallowing by releasing it momentarily and grasping it again.

The palaeoniscoids, from which the holosteans and teleosts
evolved, seem to have had a very simple jaw action.[82] Only the
lower jaw moved; the premaxilla and maxilla were sutured to
the bones of the cheek. Consequently, their mouths must have
opened in a grin. At the holostean stage of evolution the maxilla
became detached from the cheek bones, and so movable. Figs.
10A, B show how it must have moved in typical holosteans and
how it does move in the modern holostean *Amia* and in most of
the more primitive teleosts. The figure is based on the trout,
Salmo. In such fish, the anterior (dorsal) end of the maxilla bears
a spike which points towards the mid-line. This lies in a cylin-
drical cavity between the premaxilla anteriorly, and the cranium
and suspensorium posteriorly. This forms a hinge joint; the
maxilla can swing up and down about the axis of the spike. The
other end of the maxilla is attached to the lower jaw by the tough
connective tissue of the lip. As the jaw is depressed to open the
mouth, the lip pulls on the maxilla, swinging it forwards (fig.
10B). As it is raised again, the lip pulls the maxilla back. The
mechanism is similar to the mechanism of the model shown in
fig. 12G. This model consists of rigid rods, hinged together: ad
represents the lower jaw, cd the lip, bc the maxilla and ab the
rest of the skull. If ab is fixed in position and ad is moved, bc
and cd must move in a particular and predictable way. The parts
of the model have only one degree of freedom of relative movement[14] (the term 'degree of freedom' was introduced in Chapter

2). The jaws of the fish are not quite as simple as this model of them, because their lips are flexible. The lip of *Salmo* is more or less taut when the mouth is wide open and when it is closed, but it becomes slack at intermediate positions of the lower jaw and allows the maxilla some play.

An effect of the movement of the maxilla is to fill in, in part, the corner of the open mouth, making it less like the grinning mouth of fig. 12E and more like the round mouth of fig. 12F. It thus makes it easier for the fish to suck food into its mouth. The extent of the movement, and so the degree of improvement for sucking, vary from species to species. When the maxilla is short and the lip attaches it to a point well anterior on the lower jaw, it swings through a large angle, but when it is long and attached near the articulation of the lower jaw, it swings through a small one.

A more perfectly round mouth can be achieved if the premaxilla moves as well as the maxilla. In some teleosts such as the herring (*Clupea*) the premaxilla is attached to the maxilla and swings with it. In others such as the cod and its relatives (*Gadus*) there is a different arrangement which must be described in some detail because of its bearing on the evolution of the protrusible jaws which will be described in the next section of this chapter. Each maxilla has a large condyle which articulates with the cranium, and the spike which formed the articulation in *Salmo* is reduced. The premaxillae are as long as the maxillae (figs. 13A, B). They are firmly attached to a piece of cartilage, the rostral cartilage, which rests on the anterior end of the cranium. Ligaments from the cranium, suspensoria and maxillae hold the premaxillae and rostral cartilage in place, but allow the cartilage to rock on the cranium. When the mouth opens, the posterior ends of both the maxillae and the premaxillae swing forwards. The maxillae rotate about their articulation with the cranium and the premaxillae rock with the rostral cartilage. High speed films of *Gadus* feeding show that in addition to these movements there is slight protrusion of the premaxillae.

Protrusible jaws

Several groups of teleosts have evolved protrusible premaxillae, which do not merely swing forwards as the mouth opens, but

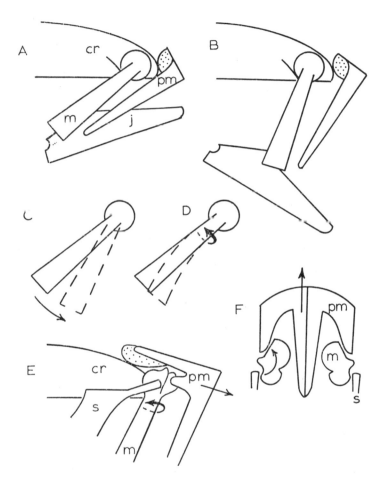

Fig. 13 The jaw mechanisms of *Gadus* and *Perca*. (A), (B) Diagrams showing the positions of the jaw bones of *Gadus* with the mouth closed and open. (C) to (F) Diagrams illustrating the mechanism of protrusion in *Perca*. *cr*, cranium; *j*, lower jaw; *m*, maxilla; *pm*, premaxilla; *s*, suspensorium. The rostral cartilage is stippled

D

slide forward over the cranium as well. The most important of these groups are the acanthopterygians,[13] the cyprinoids[12] and the cyprinodontoids or tooth-carps.[14] Together they include a very large proportion of living teleosts. The Acanthopterygii are much the largest superorder of fish and the Cyprinoidei include a large proportion of the freshwater fish of the world. The angelfish is an acanthopterygian. Figs. 12A, B show how its premaxillae are protruded as its mouth opens. The carp (*Cyprinus*) shown in fig. 1 has its mouth open and its premaxillae protruded.

The mechanisms of protrusion vary from group to group. All seem to have evolved from ancestors with long premaxillae which rocked on a rostral cartilage, like those of *Gadus*, though *Gadus* is not at all closely related to any of them. The jaw mechanism of the perch (*Perca*) will be described because *Perca* is a typical acanthopterygian, and the acanthopterygians are the largest group of fish with protrusible jaws.

The jaw bones of *Perca* are very like those of *Gadus* but the premaxillae have ascending processes which point posteriorly (fig. 13E). The rostral cartilage is attached to the ventral surfaces of the ascending processes and therefore lies well back on the dorsal surface of the cranium when the premaxillae are retracted, and does not slide off the cranium when they are protruded. When the mouth is opened widely the rostral cartilage and premaxillae slide forward through about 5% of the length of the head. In many acanthopterygians they are protruded further. In the angelfish, which has longer ascending processes, they are protruded through about 15% of the length of the head (fig. 12B), and in some species they are protruded further even than this.

The mechanism of protrusion depends on movements of the maxilla, which has a complicated head at its anterior end. It has a condyle which articulates with the cranium, as in *Gadus*, but this condyle is free to slide along the surface of the cranium as well as to rock on it. The lateral face of the head of the maxilla bears a ridge which is embedded in the skin and attached by tough dermis to the superficial bones of the cheek. When the condyle slides forward on the cranium, the ridge is held back and the maxilla rotates about an axis through this ridge. The maxilla can therefore make two types of movement. It can swing forward and back as shown in fig. 13C and it can twist about the axis through the ridge as in fig. 13D.

The ventral end of the maxilla is attached to the lower jaw by tough connective tissue, as in *Salmo* and *Gadus*. When the mouth opens, the ventral end of the maxilla must move forward. This movement can occur in either of two ways. The maxilla may swing forward as in fig. 13C or it may twist as in fig. 13D. The latter movement moves the ventral end of the maxilla anteriorly because the maxilla is curved (convex laterally).

The twisting movement causes protrusion but the swinging movement does not. The anterior end of the head of the maxilla lies deep to a process of the premaxilla (figs. 13E, F). This arrangement may have evolved originally to help hold the premaxilla firmly, for it is present in *Gadus* and various other fish whose jaws are barely protrusible. In acanthopterygians, it is responsible for protrusion. When the maxillae twist as in fig. 13D, the heads of the maxillae press against the premaxillae, forcing them forward (fig. 13F).

Both types of movement of the maxillae are limited by ligaments. The premaxillae cannot be protruded through more than a certain distance and the maxillae cannot swing forward through more than a certain angle. When the mouth is opened as wide as possible both types of movement must occur and the premaxillae are necessarily fully protruded. When the lower jaw is raised as far as possible it presses on the premaxillae and they are necessarily fully retracted. At all intermediate positions of the lower jaw, a range of degrees of protrusion is possible. Whereas the jaws of primitive teleosts have one degree of freedom of relative movement of their parts, the jaws of *Perca* have two.

There is another important mechanism which affects protrusion. The palatine bone projects from the anterior end of the suspensorium and lies lateral to the posterior part of the head of the maxilla (fig. 13E). It lies dorsal to the axis of the hinge joints between the suspensorium and the cranium, and therefore moves medially when the rest of the suspensorium swings laterally to enlarge the buccal cavity. When the buccal cavity expands the palatines press on the heads of the maxillae, tending to cause the twisting movement of fig. 13D, and so protrusion. If the mouth closes with the buccal cavity expanded, the premaxillae remain protruded.

The mechanisms of the jaws can easily be demonstrated on a fresh dead specimen of a perch or of any other typical acanthopterygian. Preserved specimens are too stiff. The halibut

(*Hippoglossus*) is quite suitable, although its jaws are asymmetrical, and can be obtained easily from British fishmongers. If the mouth of a fresh dead perch or halibut is opened widely by depressing the lower jaw, the premaxillae are protruded. The buccal cavity can be expanded by pulling the tissue which connects the ventral ends of the pectoral girdle to the hyoid bars; this is muscle (v) of fig. 10D. If the mouth is closed with the buccal cavity expanded, the premaxillae remain protruded. Fig. 12C shows an angelfish in this position. The mouth has closed after sucking in food, with the buccal cavity still expanded, and the premaxillae remain fully protruded.

In cyprinoids and toothcarps the mechanisms of protrusion are quite different and seem to have been evolved independently of acanthopterygians and of each other.[14] Further, the premaxillae are not locked in the protruded position by the palatines when the buccal cavity is expanded. Instead, they can be held protruded by contraction of a special muscle which has differentiated from the adductor mandibulae. In *Gadus* and in *Perca* parts of the adductor mandibulae insert on the maxilla instead of on the lower jaw, but apparently merely help to swing the maxilla posteriorly (i.e. to reverse the movement of fig. 13C) when the mouth closes. Cyprinoids and toothcarps have similar muscles, but they have the additional function of protruding the premaxillae.

In spite of the mechanical differences, cyprinoids and toothcarps move their jaws in very much the same way as acanthopterygians. The premaxillae are protruded as the mouth opens to take food. They remain protruded (or may even be protruded further) as the mouth closes again, except perhaps in toothcarps when they are feeding at the surface.

It has often been written that the advantage of protrusion is that it moves the mouth suddenly nearer the prey at the last moment before it is captured, and so gives the prey less chance of escaping. This advantage seems likely to be small in most cases. Numerous teleosts with protrusible premaxillae eat slow-moving and stationary food, which does not need a particularly rapid approach. Predators chasing fast-swimming prey seem unlikely to be helped much by sudden protrusion of the premaxillae through a mere 10% or so of the length of the head. One of the few fish to which this advantage seems likely to be important is the John Dory. It can protrude its premaxillae further than most other acanthopterygians, through about a quarter of the length of the head. We have already seen how it approaches its prey stealthily and sucks it in suddenly. As it is presumably the last stages of stalking the prey that are the most difficult, this protrusion may well be useful in getting the mouth opening near enough to the prey to suck it in.

A large proportion of fish with protrusible premaxillae take at least part of their food from the bottom. They may obtain a different advantage from protrusion.[13] The following suggestion depends on the assumption that it is an advantage, when feeding from the bottom, to have the axis of the body as nearly horizontal as possible. A fish must return to the horizontal before it can flee from the predator, and river fish probably find it difficult to maintain an oblique stance in flowing water.

When the mouth of a typical teleost is closed, the lower jaw slopes upwards from the articulation. It could hardly be arranged otherwise in a fish with a streamlined body and a terminal mouth. Consequently, the anterior end of the lower jaw swings forward as the mouth opens. If the upper jaw is not protruded, the open mouth will point somewhat upwards. The open mouth of the angelfish in fig. 12B points almost directly anteriorly. If the premaxillae had not been protruded, it would point upwards at an angle of about 30° to the long axis of the body, and the fish would have to approach the food at a correspondingly steeper angle to get the dorsal edge of the mouth as near to it as in fig. 12B. A really close approach of the mouth to the food is important if the estimate is correct, that the greatest distance from which food can be sucked is about a quarter of the length of the head.

The forward swing of the jaw as the mouth opens is particularly marked in fish with deep bodies like the angelfish, in which it slopes steeply when the mouth is closed. It is exaggerated in fish which swing their crania upwards as they open their mouths. It is cancelled out in some fish which swing their lower jaws down beyond the horizontal, as *Gadus* does (fig. 13B). The mouth of *Gadus* points almost directly anteriorly when it is opened widely although the premaxillae are only slightly protruded.

A mouth that points anteriorly is not necessarily ideal. Mouths that point upwards are common among tropical fresh-water fish which feed largely on insects caught at the surface, and among fish which, like *Lophius*, rest on the bottom and catch fish swimming above them. Many fish which feed mainly from the bottom have mouths which point downwards. The gudgeon (*Gobio*, Cyprinoidei) feeds almost entirely from the bottoms of the streams in which it lives. Its main food is chironomid larvae.[50] Its mouth is terminal when it is closed with the premaxillae retracted, but the premaxillae are strongly protrusible so that it can be pointed at an angle ventrally. The body is rather shallow, as in most bottom-living fish, and is consequently rather flexible dorso-ventrally. By bending down its head and protruding its pre-maxillae the gudgeon can point its mouth vertically downwards while most of its body is horizontal. It swims along in this position, applying the mouth closely to the bottom like the nozzle of a vacuum cleaner, and periodically opening it to suck food in.[12]

So far we have considered only the advantages of protruding the premaxillae as the mouth opens. Now we must consider the possible advantages of keeping them protruded as the mouth closes. One advantage is probably the effect on the speed of closing the mouth. If the premaxillae were retracted, the lower jaw would have to be raised further to close the mouth, and this would presumably take longer. Keeping the premaxillae pro-truded may also increase the volume of the buccal cavity with the mouth closed, and so the volume of water that can be sucked into the mouth without being blown out again as the mouth closes. Rough measurements on a cyprinoid indicate that this might give a useful advantage, but in acanthopterygians its effect seems to be very slight.[13]

Food is not always taken into the mouth by a single suck. It may be taken in part way and grasped between the jaws while the buccal and opercular-cavities are collapsed in preparation for a second sucking action. When this is done, the ability to move the premaxillae and lower jaws independently may be useful, because it may enable the fish to adjust the angle of the food before sucking it in. It seems likely that swallowing may be easier if the prey is pointed towards the gullet.

In most fish with protrusible premaxillae, the lower jaw simply swings about its articulation with the suspensorium as the mouth opens. In the exceptional case of Epibulus, the part of the sus-pensorium which bears the articulation swings forward as the mouth opens, and the lower jaw moves bodily forward as it swings down.[6] This, with the protrusion of the premaxillae, produces a long tapering snout which Epibulus inserts into crannies in the coral reefs where it lives, to catch small crusta-ceans.[53]

Teeth

The teeth on the jaws of bony fish are usually conical, with a sharp point. Their most usual function is for holding prey and they are often inclined posteriorly, which gives the prey less chance of escaping. *Lophius* and some other teleosts have large teeth which are not rigidly fixed, but are attached to the underlying bone by ligaments on the inner sides only of their bases. These ligaments are elastic and allow the teeth to fold down, pointing into the mouth, as prey enters. The teeth cannot be bent in the opposite direction, so it is much harder to escape from the jaws than to get into them. The only easy way out of the mouth of an angler fish is into its gullet.

Teeth used for grasping vary greatly in size. Some fish-eating teleosts such as the pike have relatively small numbers of large pointed teeth. Other equally voracious ones such as the Nile perch (*Lates*) have very large numbers of tiny teeth arranged in broad bands. These teeth are too small to pierce the skin of the prey but make admirable friction pads for holding it. The mouths of pike and of most of the other predators with large teeth open in a grin, as in fig. 12E, whereas the mouths of Nile perch and of many other predators with small teeth are round, as in fig. 12F.

Large teeth probably give more certainty of grip, but they have two disadvantages. First, they make it necessary to open the mouth wider to admit prey, unless they are hinged as in *Lophius*. If the mouth is opened wider, the speed at which water can be drawn in is, as we have seen, less. Perhaps this is why mouths which are round, which is the shape best adapted for sucking, tend to be provided with small teeth. Secondly, large teeth damage the prey and may thereby release substances into the water which may warn other potential prey. This is likely to be particularly important in fresh water, because a very large proportion of freshwater fish are Ostariophysi. Pfeiffer[79] has shown that most Ostariophysi have alarm substances in their skin, which are released when they are wounded. The odour of the alarm substance induces a fright reaction in members of the same species, and even in other Ostariophysi. The form of the fright reaction varies between species, but it is always such as to aid escape from a predator.

Though their usual function is grasping, fish teeth are sometimes specialised for cutting. The Characinoidei are common in fresh water in tropical Africa and South America. Many of them have large teeth with cutting edges instead of simple points.[8] In the more primitive ones the edge can be seen to consist of a row of subsidiary cusps on either side of the main one. The cutting teeth of the lower jaw fit closely inside those of the upper jaw, so that the two rows of teeth cut like scissors when the mouth closes. *Myleus* is a herbivorous characin. The gut of a specimen which I opened was filled with pieces of leaves which had apparently been bitten from underwater plants. The pieces had cleanly cut serrated edges which matched the line of the cutting edges of the teeth. The piranhas (*Serrasalmus*, etc.) are carnivorous characins. They have the unusual habit of biting pieces from fish which are too large to swallow whole, and sometimes attack mammals which enter the South American rivers where they live. The teeth of *Myleus* are chisel-like in section, with the faces meeting at quite a large angle at the cutting edge. Those of *Serrasalmus* are much more slender and knife-like. They are much better suited for penetrating deeply into flesh, though the teeth of *Myleus* are entirely suitable for cutting thin sheets such as leaves. *Serrasalmus* has enormous jaw muscles whose accommodation has required considerable enlargement of the head and some curious anatomical rearrangement within it (Chapter 7).

Though few teleosts bite pieces from fish, many bite pieces from invertebrates. The plaice (*Pleuronectes*) has a single row of closely spaced chisel-shaped teeth in each jaw. Their cutting edges close against each other, like the cutting edges of a pair of pincers and not like scissors. Plaice sometimes bite off the siphons of lamellibranchs and the anterior ends of tubicolous polychaetes, which are the only parts of these animals which are accessible when they are buried in the sand or mud. Plaice also capture complete lamellibranchs, and swallow them with their shells. Some fish which live on coral reefs bite off individual polyps of coral, leaving the hard skeleton of the coral untouched. They include some of the butterfly fishes (Chaetodontidae). which have small mouths with protruding teeth on the ends of long snouts.[53]

The puffer fishes (Tetraodontidae) have their teeth fused into powerful beaks. They bite off and swallow hunks of hard coral, and digest the small proportion of organic matter which they contain.[53] Just inside the cutting edge of the beak, in each jaw, is a convex plate which is apparently used for crushing molluscs. Crushed molluscs are found in the gut as well as pieces of coral and various other foods. Crushing probably speeds the digestion of molluscs, by admitting enzymes to the tissues inside the shell. The shanny, *Blennius*, has strong teeth rather like those of plaice. It feeds mainly on barnacles which it breaks off rocks. Where corals and barnacles occur, they are often plentiful, but only a few specialised fish such as the puffer fishes and the shanny are able to feed on them.

Many submerged surfaces are covered by a film of algae, and many teleosts feed on this film. Among them, the sucking catfish (*Plecostomus*) is quite a popular fish in tropical aquaria. It has a ventral mouth surrounded by a sucker which it uses to anchor itself to stones and to the walls of aquaria.[10] Its teeth are long and flexible, like bristles, and curved inward at their ends. Unlike most other catfish, it has movable premaxillae; the muscles that move them seem to have differentiated from the muscles which, in other catfish, move the sensory barbels. Its natural food is algae growing on rocks in streams, but aquarium specimens can often be seen feeding on algae growing on the walls of the aquarium. They move slowly along, opening and closing the mouth by movements of the premaxillae and lower jaw, and so brushing the glass with their bristle-like teeth. They do not seem to be able to use the sucker while they are feeding. Since the teeth are flexible, they will accommodate themselves to irregularities in the surface which they are brushing. However, teeth are not necessary for scraping up algae, and a number of cyprinoids use toothless horny jaws for the purpose. *Gyrinocheilus* and *Labeo*[40] are examples. *Gyrinocheilus* has a sucker rather like that of *Plecostomus*, but *Labeo* does not.

The teeth of bony fish are not limited to the jaws. There are often teeth on the buccal surfaces of the cranium and suspensorium, and on bony plates attached or fused to the buccal faces of the gill skeleton. The latter are called pharyngeal teeth, and seem to be used mainly for swallowing. There is usually a group of pharyngeal tooth plates attached to the dorsal ends of the gill arches, and a plate attached to the lower half of the last branchial arch (posterior to the last gill slit) on each side of the body (fig. 9A). The teeth on these plates are usually pointed and inclined posteriorly. If the food is held between the upper and lower pharyngeal teeth, and the lower ones are moved anteriorly and posteriorly, the food will be moved posteriorly by a ratchet effect into the gullet.

Various teleosts including the plaice have large blunt pharyngeal teeth which they seem to use for crushing molluscs. In Lake Victoria, the cichlid *Astatoreochromis* feeds almost entirely on molluscs whose shells are found crushed in its stomach.[44] They are presumably crushed by its pharyngeal teeth, which are large and blunt. In the other lakes in which it lives, the same species eats insects, plants and small fish, but few molluscs, and has considerably smaller teeth and toothplates. *Astatoreochromis* from Lake Victoria have been kept and bred in New York, where they have been given very few molluscs to eat. A specimen that was examined after a few generations was found to have pharyngeal teeth and toothplates which were most unlike those of its Lake Victoria ancestors, and much more like those of specimens from the lakes where they eat few molluscs. It appears that feeding on molluscs has in this species an effect on the development of the apparatus which is used to crush them. It may be that their development is promoted by calcium obtained from ingested shells, or there may be a more direct effect of use on their development. Many cases are known of mechanical factors influencing the size and shape of mammal bones.

The cyprinoids have no teeth on their jaws, but have characteristic lower pharyngeal teeth.[73] There are no upper pharyngeal teeth but their place is taken in most species by a tough horny pad attached to a projection of the skull. The lower toothplates are fused to the underlying bones (the last ceratobranchials) and bear large teeth which point medially and are hooked upwards at their ends (fig. 14A–D). The dorsal edges of the teeth are often fairly sharp and serrated (fig. 14B). The teeth alternate on the two ceratobranchials, so that they inter-digitate (fig. 14D). Many muscles insert on the ceratobranchials but the largest are the two which are described below. The movements they cause can be demonstrated by stimulating them electrically in a fish which has just been killed by a blow on the head. I have used orfe (*Idus*) for this purpose.

The muscle which is labelled (i) in fig. 14D originates in a huge cavity in the cranium. It pulls the ceratobranchial dorsally and laterally. If the two ceratobranchials are moved simultaneously in this way, food in the pharynx will be pressed against the horny pad and torn by the opposite movements of the hooks on the two sets of teeth. Living fish seem to do this. If orfe are fed with pieces of earthworm, they make vigorous gasping movements for

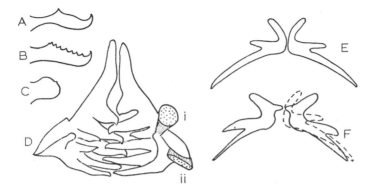

Fig. 14   (A) to (D) Pharyngeal teeth of Cyprinidae. Posterior views of single teeth of (A) orfe (*Idus*), (B) rudd (*Scardinius*) and (C) carp (*Cyprinus*). (D) Dorsal view of the pharyngeal teeth of *Idus* with their supporting bones and principal muscles. (E), (F) Occlusal views of the toothplates of the African lungfish (*Protopterus*). (E) Upper toothplates and (F) lower toothplates with one upper one superimposed in the position it occupies when the mouth is closed

some minutes after taking them into their mouths. It is of course impossible to see what is going on inside the pharynx, but if the orfe are killed and the pieces of worm recovered, the pieces are found to be lacerated, with the contents of the coelom exposed through the torn body wall. This sort of treatment probably speeds digestion by making more of the food immediately accessible to enzymes.

The muscle labelled (ii) in fig. 14D originates on a further posterior extension of the process of the skull which supports the horny pad. It pulls the ceratobranchial dorsally and posteriorly, scraping the edges of the teeth along the horny pad. This movement is presumably used in swallowing, but may also serve to grind food. Other smaller muscles connect the ceratobranchials to each other and to the pectoral girdle, and must serve as the antagonists of muscles (i) and (ii).

The hooked teeth shown in figs. 14A, B are typical of cyprinoids such as the orfe which take a mixed diet consisting largely of insect larvae and plant material. The common carp (*Cyprinus*, fig. 1) eats a similar mixture of foods but has simple knob-like pharyngeal teeth without hooks or serrations (fig 14C). They become heavily worn, and the striations on the worn surfaces are longitudinal. This indicates that muscle (ii) is used more than muscle (i) in mastication. The Chinese grass carp (*Ctenopharyngodon*) is almost exclusively vegetarian. It feeds on the leaves of aquatic plants, and of terrestrial plants submerged by flooding. Its pharyngeal teeth have reduced hooks, blunt dorsal edges and very marked serrations. Plant material in the gut is found in fragments whose dimensions match the spacing of the teeth, suggesting that they have been produced by the tearing action attributed to muscle (i).[54] It can be seen under the microscope that superficial cells have been scraped off the fragments, perhaps by the action of muscle (ii). Fish have no enzyme which will break down plant cell walls, but *Ctenopharyngodon* is able to extract about half the protein from its masticated food.

Since the pharyngeal teeth of cyprinoids lie posterior to the gills, mastication need not interrupt respiration.

The Dipnoi are peculiar in having no premaxillae or maxillae, and no real teeth. They have ridges of dentine-like tissue on the buccal surface of the cranium and suspensorium which interdigitate beautifully with similar ridges on the lower jaw (fig. 14E).[77a] The African lungfish (*Protopterus*) feeds mainly on molluscs which are found in the gut with their shells crushed into small fragments.[34]

Gill rakers

When the food is small it must be prevented from escaping from the buccal cavity through the gills. The gill rakers do this. They are processes which project from the gill arches like the teeth of a comb. Often each gill arch bears two rows of gill rakers, as shown in fig. 9B, which between them fill the spaces between one arch and the next. Predators which take large food usually have widely spaced gill rakers, but fish which eat small food have very fine, closely spaced ones. Sometimes the gill rakers are used as a sieve to separate the food from other material, and not merely as a means of retaining it. *Lethrinops* is a cichlid living in Lake Nyasa which feeds by filling its mouth with sand which is then discharged through the opercula.[40] The gill rakers strain out burrowing invertebrates. Burrowing chironomid larvae form the main stomach contents of a species with rather widely spaced gill rakers,

while ostracods are the main food of a species with more closely
spaced ones. The way in which the anchovy uses its gill rakers to
strain plankton from water has already been described.

Adaptive radiation

A good deal of this chapter has been concerned with adaptations
of teleosts to particular diets. Such adaptations can be seen
particularly clearly when a number of closely related fish are
adapted to different feeding habits. Some of the most remarkable examples of adaptive radiation are provided by the cichlids
of the East African lakes.[43] In Lake Victoria, for instance, there
are about 120 species of *Haplochromis*, most of which occur
nowhere else, adapted to at least five clearly distinct diets. They
are believed to have evolved from a few ancestral species in the
$1-1\frac{1}{2}$ million years since the lake was formed. Some, which are
believed to be most like the ancestral stock, feed on insect larvae
and detritus. A few, which have evolved oblique scraping edges
on their teeth, scrape algae from plants and rocks. Some of the
larger species are fish-eating predators, with long jaws and strong
sharp teeth. Some species with large mouths and reduced teeth
feed on *Haplochromis* embryos and larvae. *Haplochromis* keep
their eggs, and their young for some time after hatching, in their
mouths. It is believed that the species which eat young *Haplochromis* enclose a brooding parent's mouth in their own, and
somehow force the parent to release its brood. Still other species
of *Haplochromis* feed on molluscs, and they fall into two distinct
groups. Some eat complete molluscs and crush them with their
large pharyngeal teeth. Others have relatively weak pharyngeal
teeth, but have strong hooked teeth on their jaws which they use
to grab the soft parts of crawling gastropods. They detach the
shell and swallow only the soft parts.
   A rather similar example of adaptive radiation is found in
Lake Lanao, in the Phillipines, where there are 18 species of
Cyprinidae showing diverse specialisations.[76] They all seem to
have evolved from a single (identifiable) ancestral species in the
10,000 years or so since the lake became isolated. Another
remarkable radiation is found among the Characinoidei of the
rivers and swamps of tropical South America.[8] There are about
200 genera, and some of them have highly specialised teeth and
jaw muscles. *Serrasalmus*, which bites pieces from fishes, and

Myleus, which bites pieces from leaves, have already been referred to. There are also predatory genera with long jaws and large teeth which swallow fish whole, small insectivorous genera, and genera with protrusible jaws and tiny teeth which feed on detritus. There are other genera with peculiarities which have no direct bearing on feeding; for instance, *Erythrinus* is able to survive in stagnant swamps because it can breathe air (Chapter 4). The flatfish (Pleuronectiformes) provide a good example of adaptive radiation to various feeding habits in the sea. The halibut and plaice, for instance, resemble each other in so many respects that it seems clear that they are closely related and they are put in the same family, but they have very different feeding habits. The halibut feeds on active prey such as fish and shrimps. It has a large almost symmetrical mouth and has sharply pointed teeth on its pharyngeal tooth plates. The plaice feeds largely on molluscs and tubicolous polychaetes. It has a small mouth, strongly inclined to the side of the fish which lies on the bottom. It has chisel-like teeth on its jaws which it uses to bite pieces from its food, and its pharyngeal toothplates have blunt teeth and large muscles and are used for crushing mollusc shells.

Other fish

The rest of this chapter is about fish which are not Osteichthyes.

The Agnatha have no jaws, so their feeding mechanisms are very different from those of other fish. The hagfishes have a pair of toothed plates which can be protruded from the mouth to grab food and draw it in, rather like the paragnaths of polychaete worms.[30] They eat a wide variety of foods. The ammocoete larvae of lampreys are filter feeders, which use a single current of water to supply their oxygen and their food. They pump this current in through the mouth and out through the gills, filtering it through a sheet of mucus in the buccal cavity. Minute organisms in the water are trapped by the mucus, which is drawn slowly into

Fig. 15 The jaws of a selachian. (A) to (C) Tracings of photographs of the head of a piked dogfish (*Squalus*) in three positions which are described in the text. (D) Lateral view of the skull of *Squalus*, with the position of the adductor mandibulae indicated by a broken arrow. *cr*, cranium; *hm*, hyomandibular cartilage; *mc*, Meckel's cartilage; *pq*, palatoquadrate cartilage

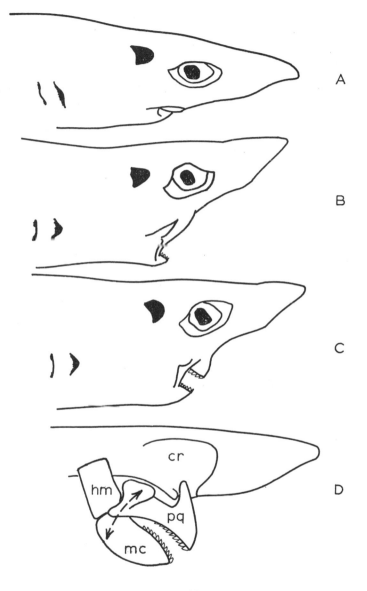

Fig. 15

the gut. Adult lampreys have suckers round their mouths which they use to anchor themselves to stones. The respiratory system has been rearranged to make simultaneous suction and respiration possible; the transition from ammocoete to adult lamprey involves almost as remarkable a metamorphosis as the transition from tadpole to adult frog. The adults of some species of lamprey also use their suckers to attach themselves to other fish, whose blood they suck.[2]

The Holocephali have very hard ridges on their jaws, rather like the toothplates of Dipnoi.[77a] *Chimaera* eats a variety of invertebrates which are found in small pieces in the gut.[3] They are presumably broken up by the toothplates.

The skull of a selachian is shown in fig. 15D. It is made of cartilage and is in many respects very different from the teleost skull shown in Fig. 10. The cartilages are homologous with the various cartilages which are formed in teleost embryos and later converted to bone, but there are no homologues of the dermal bones which are. formed without a cartilage precursor. There are no premaxillae or maxillae. The lower jaw is formed entirely by Meckel's cartilage, which forms only a small part of the lower jaw in teleosts. The upper jaw is formed by the palatoquadrate cartilage which is incorporated in the suspensorium in teleosts. The posterior end of this cartilage is attached by ligaments to the hyomandibular and Meckel's cartilages. The hyomandibular cartilage is homologous with the part of the suspensorium which forms the posterior hinge joint with the cranium in teleosts. It forms a sloping hinge joint with the cranium: the anterior end of the joint is dorsal to the posterior end. The palatoquadrate cartilage has a dorsal process which rests in and is free to slide along a groove in the cranium. It can slide anteriorly and ventrally, or posteriorly and dorsally.

I have no photographs of selachians feeding but have traced figs 15A–C from photographs of a fresh, dead piked dogfish (*Squalus*). They show the jaws in three positions which are probably adopted, in the same order, in feeding. They are comparable to the three positions of the angelfish in figs. 12A–C. In fig. 15A the mouth is closed and the buccal cavity is contracted. The mouth, which is ventral as in most other selachians, lies flush with the undersurface of the head, and the whole head seems well streamlined. In this position, the hyomandibular cartilages project almost horizontally from the cranium. In fig.

15B the mouth is wide open. The upper lip has swung down, filling the corner of the mouth, much as the maxillae of teleosts swing forward as the mouth opens. The buccal cavity has enlarged considerably, but the hyomandibular and palatoquadrate cartilages are in more or less the same positions as in fig. 15A. In fig. 15C the mouth is almost closed with the buccal cavity still expanded, as it would presumably be immediately after taking food. The hyomandibular cartilages have swung ventrally and anteriorly about their sloping hinge joints with the cranium. They have pushed the palatoquadrate cartilages forward and down so that the upper jaw projects from the ventral surface of the head. This is the only way in which the mouth can be closed with the buccal cavity expanded. The position is curiously similar to the position of teleosts immediately after taking food with the mouth closed and the premaxillae (if protrusible) protruded. The dogfish (*Scyliorhinus*) cannot protrude its upper jaw so far.

The adductor mandibulae runs from the palatoquadrate to Meckel's cartilage (fig. 15D). Another muscle runs from the cranium, anterior to the eye, to the fascia of the adductor mandibulae or to the upper lip (to the latter in *Squalus*). It may contract as the mouth closes and aid protrusion of the jaws.

Some of the larger sharks are notorious for their habit of biting pieces from large prey. If the upper jaw always ended flush with the ventral surface of the head, it would be difficult for them to bite things because the surface of the head would often get in the way. If the jaws always protruded from the head, they would increase the drag on the fish by interfering with its streamlining. The protrusible jaws of sharks make it easier for them to bite things, but leave their heads nicely streamlined when they are not feeding.

The teeth of the dogfish, *Scyliorhinus*, are small and very like the placoid scales which cover its skin. Their points point into the mouth and they hold the prey rather than cut it. Dogfish feed on small teleosts and various invertebrates. Remains of a large gastropod mollusc, *Buccinum*, are often found in their guts, but these remains never include the shell. A dogfish has been seen attacking a *Buccinum* which was crawling in its aquarium. It knocked the *Buccinum* over with its snout and grabbed its soft parts. It then shook its head from side to side until its mouthful tore away from the rest of the mollusc.[29]

The sharks which bite pieces from large prey have much larger and more specialised teeth. Often the teeth on one jaw are broad triangles with sharp cutting edges while those on the other are much more slender, with sharp points. The triangular teeth cut the prey while the pointed ones hold it. The shark grasps the prey and makes vigorous rolling movements by moving its pectoral fins alternately.[86] The inertia of the prey and the resistance of the water prevent it from twisting with the shark, so the cutting teeth move from side to side through the prey like the teeth of a saw. Some sharks get the same effect by shaking their heads from side to side.

Mustelus feeds largely on crabs and similar Crustacea whose shells might damage sharp teeth but can be gripped and crushed by blunt ones.[3] It has flat square teeth arranged like paving-stones on its jaws. The eagle-rays (*Myliobatis*, etc.) feed largely on molluscs and have very strong pavement-like plates of teeth to crush them.[6]

6

SOME SENSE ORGANS

The sense organs which will be discussed in this chapter are the lateral line system and some other sense organs that are believed, on morphological and embryological grounds, to have evolved from it. These others are the semicircular canals and otolith organs of the ear, and various electric receptors which are sometimes referred to as the ampullary lateral line organs. The lateral line system and its derivatives are often referred to together as the acoustico-lateralis system.

Good sense organs can obviously give a selective advantage. A fish with better sense organs than its fellows may be able to find more food, or may use less energy in finding food. It may get earlier warning of the approach of predators or may be able to find its way more quickly when escaping from them. It may be better able to co-ordinate its movements and this may result both in saving of energy and in improved ability to catch prey and escape predators. Some of the organs of the acoustico-lateralis system serve to detect objects such as food, predators and rocks, some help in the co-ordination of movement and some do both.

Neuromasts

The lateral line system serves to detect disturbances in the water. Its individual sense organs are neuromasts (fig. 16A).[36] In their simplest form each consists of a lump of jelly, known as the cupula, which rests on a sensory epithelium and projects from the surface of the fish. The sensory cells are known as hair cells because each bears a hair-like bunch of fine processes which extends into the cupula. Most of the processes in the bunch are long microvilli of the kind sometimes called stereocilia; they show little internal structure when sections are examined by electron microscopy. One process at the side of the bunch has the structure of a true cilium or kinocilium. It contains a pair of filaments running up its centre and nine double filaments arranged round them. Not only is the kinocilium always at the edge of the bunch, but it is always oriented in the same way (fig. 16B). Its two central filaments lie in a plane perpendicular to the radius of the bunch. It has a basal body, as ordinary cilia do, and this basal body has a projection which points radially away from the bunch. The cupula seems to be attached to the epithelium only by the hairs of the hair cells. Sensory neurones synapse with the bodies of the hair cells. These neurones are unusual in that a succession of action potentials travels along them even when no stimulus is applied to the neuromast.

When a current of water impinges on the side of a cupula, it has two effects. It makes the cupula bend, and it makes it slide along the epithelium. The sliding movement affects the frequency of action potentials in the neurones if its direction is right.

At this point it is convenient to leave the neuromasts of the lateral line for a while and examine the very similar neuromasts of the semicircular canals of the ear, which happen to have the kinocilium on the same side of the hair of every hair cell.[39] Because of this, it is possible to establish the relationship between

---

Fig. 16 Diagrams of various parts of the acoustico-lateralis system. (A) A lateral line neuromast showing the cupula (*cu*), some hair cells (*hc*) and nerve endings (*n*). (B) Section through a single sensory hair showing the kinocilium (*k*) and stereocilia (*s*). The two central filaments of the kinocilium are shown, and the projection of its basal body is indicated by a broken line. (C) A semicircular canal. (D) A superficial neuromast and one in a canal which opens to the surface by the pores *a* and *b*. The arrows represent water movements and are discussed in the text. (E) Transverse section through the swimbladder of a fish with a Weberian apparatus, showing how expansion of the swimbladder widens the slit in the tunica externa. (F) Dorsal view of the Weberian apparatus, not to a consistent scale: *os*, Weberian ossicles; *sa*, sacculus with its otolith; *te*, tunica externa; *ti*, tunica interna seen through the slit in the tunica externa; *vc*, vertebral column, which runs dorsal to the swimbladder

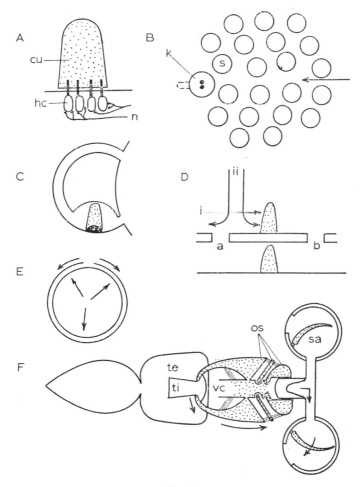

Fig. 16

the direction of sliding, the orientation of the hair cells and the frequency of the action potentials.

The semicircular canals of fish are just like the semicircular canals of mammals, whose structure and function are familiar to most biologists. One is represented in fig. 16C. It is a tube opening at both ends into the part of the ear known as the utriculus. Canal and utriculus are filled with fluid. Part of the canal is swollen to form the ampulla which contains the neuromast. The sensory epithelium forms a hillock on which the cupula rests, and the cupula almost fills the cross-section of the ampulla. If the fluid in the canal moves relative to the wall of the canal, the cupula must move with it. Deflections can be caused by rotation of the animal. If a canal is rotated clockwise, the fluid in it lags behind on account of its inertia, and the cupula is deflected anticlockwise. When rotation ends the cupula is returned to its undeflected position by elastic forces. The cupula is very delicate and transparent, but can be made visible through the wall of the semicircular canal by vital staining. The movements of stained cupulae of live pike have been watched and photographed.

Of the three semicircular canals in each ear, one has its kinocilia on the side of the hair cells nearest the utriculus. The other two have them on the side furthest from the utriculus. Action potentials have been recorded from the nerves of the semicircular canals of selachians, in apparatus in which an ear could be rotated.[1] When the ear was stationary, a regular discharge of action potentials was usually found. When it was rotated in such a way as to deflect the cupula of the canal in question in one direction the frequency of action potentials increased, but when it was rotated in the opposite direction the frequency decreased. The increase occurred when the deflection of the cupula was towards the side on which the kinocilia of the hair cells lay (i.e. in the direction indicated by the arrow in fig. 16B).

None of the other neuromasts in the ear or in the lateral line system have all their kinocilia on the same side of the hair cells. However, it seems to be a general rule that deflection towards the side on which the kinocilium lies stimulates a hair cell in such a way as to increase the frequency of action potentials in its neurone.

Because of their position in the canals, the semicircular canal cupulae can only be deflected along the line of the canal. Fish have lateral line neuromasts projecting from the surfaces of their bodies, which can be deflected in any direction. The clawed toad,

Xenopus, has similar neuromasts, and action potentials have been recorded from their neurones.[36] As in fish, half the kinocilia of each neuromast lie on one side of their hair cells and half on the other. Each neuromast is served by two neurones which carry a regular succession of action potentials when no stimulus is applied. When the cupula is moved by a microneedle or a current of water, the effect depends on the direction of movement. If it is moved so that every sensory hair is bent either towards its kinocilium or away from it, the frequency of action potentials increases in one of the neurones and decreases in the other. Movement in the opposite direction has the opposite effect. When the cupula is moved at right angles to these directions, the frequency of action potentials is affected very little, if at all. Presumably each neurone serves hair cells which all have the same orientation, and the frequency of action potentials increases when they are bent towards their kinocilia.

The relationship between the direction of movement and its effect suggests fascinating possibilities.[39] Ordinary metazoan cilia seem able to beat in only one direction. At least some of them have a projection from the basal body like the projection found in hair cells. The effective stroke in the beat is in the direction of this projection. This is at right angles to the plane joining the two central filaments of the cilium. The direction in which the kinocilium of a hair cell must be moved to increase the frequency of action potentials in its neurone is thus the same as the direction of the effective stroke of an ordinary cilium. When a neuromast is deflected, is it the kinocilia that produce the signal which leads eventually to the electrical effects in the neurones? Do they do this by reversing the normal mechanism of operation of cilia? A piezoelectric crystal can be used either to produce vibrations in response to electrical changes, or to produce electrical changes in response to vibrations.

Disturbances in water

The lateral line organs are receptors for mechanical disturbances in water. To understand them we must know something of the kinds of disturbance which can be caused by a body moving in water.

When a body moves through water the water makes way for it. This must happen whether the body is moving steadily or vibrating, but in the latter case the disturbances which are caused are periodic and are sometimes referred to as the acoustic near field of the body.

The near field is formed because the compressibility of water is limited. If water were totally incompressible it would be the only disturbance caused by a vibrating body. Since water is somewhat compressible, sound waves can be caused as well. They are sometimes described as the acoustic far field of the body.

The amplitude of near field vibrations is relatively large near a vibrating body. Water in contact with the body moves with the same amplitude as the body. However, the amplitude of near field vibrations decreases rapidly with distance from the body.[92] The amplitude of far field vibrations is relatively small close to the body, but falls off less rapidly with distance, and there is a certain distance beyond which the amplitude of the far field is greater than the amplitude of the near field. This distance is about $30,000/f$ cm from the centre of a sphere vibrating or pulsating with frequency f sec^{-1} (vibration is periodic movement backwards and forwards along an axis and pulsation is periodic increase and decrease of size).

In an acoustic field, displacements and fluctuations of pressure occur. The ratio of the amplitude of the pressure changes to the amplitude of the displacements is higher at a distance from the source where the far field is predominant than it is close to the source where the near field is predominant.

Lateral line organs

The simplest lateral line organs of fish are neuromasts projecting from the surface of the body (fig. 16A). Some experiments on the superficial neuromasts of *Xenopus* have already been described. The superficial neuromasts of fish probably work in the same way, responding to water movements which displace them in either direction along the axis marked by the kinocilia. The forces which displace them are drag forces.

In discussing the functions of these and other lateral line organs, we must consider two main types of flow. First, when a fish swims or when it rests on the bottom in a current, water flows over its whole surface, parallel to the surface. This sort of flow is indicated by the arrow (i) in fig. 16D. Superficial neuromasts can provide information about this type of flow. Secondly, animals moving in the water near the fish set up disturbances in the water round the fish. Movements parallel to the surface of the fish set up currents parallel to it and can obviously stimulate the superficial neuromasts. Movements at right angles to the fish set up currents at right angles to it, but these are deflected by the body of the fish, as indicated by the arrows (ii) in fig. 16D, and so can stimulate the superficial neuromasts.

Lateral line canals lie just below the surface of the body. Many fish have a canal running along each side of the body, and a system of branching canals on the head. Each has a series of openings to the surface and contains a neuromast between each opening and the next (fig. 16D). The cupulae extend right across the lumen of the canal, more or less blocking it and preventing flow of fluid past themselves. The hair cells are so arranged as to be sensitive to the displacements of the cupulae along the canals. The neuromast in the figure will be displaced whenever there is a difference in hydrodynamic pressure between *a* and *b*. A difference in hydrostatic pressure due to *a* being higher than *b* or vice versa will have no effect. Differences of hydrodynamic pressure arise when fluid accelerates or decelerates.[91]

The canal neuromasts can be stimulated by both the types of flow which stimulate superficial neuromasts. However, they are probably relatively insensitive to the longitudinal flow over the whole body which occurs when the fish is swimming or when it is resting in a current. As water flows over a streamlined body its velocity changes, but not very much or very rapidly, except at the front of the body, so there are no steep gradients of hydrodynamic pressure except round the head. The differences in hydrodynamic pressure between successive openings of the lateral line canals of a fish, due to this type of flow, will tend to be small except at the snout.

Disturbances due to a moving body near the fish may only cause small velocities of flow over the fish, but the pressure changes which they cause tend to be local and there may be relatively large differences between successive openings of the canals. In fig. 16D the flow (ii) results in a pressure difference between *a* and *b* and so stimulates the canal neuromast as well as the superficial one. If the flows (i) and (ii) involve equal velocities of flow over the superficial neuromast they will cause equal drag forces and stimulate it equally, but flow (ii) will involve a bigger pressure difference between *a* and *b* than flow (i) and will have more effect on the canal neuromast than flow (i).

The lateral line neuromasts can be stimulated by periodic water movements as well as by steady ones; they can be stimulated by acoustic near and far fields. The relationship between stimulus and response cannot be expected to be the same for a periodic stimulus as for a steady one. When the stimulus is steady, equilibrium is reached with the drag (on a superficial neuromast) or the pressure difference (for a canal neuromast) balanced by elastic forces in the neuromast. When the stimulus is periodic, equilibrium is not reached and the amplitude of displacement of the cupula depends on inertia and viscosity as well as on stiffness. At low frequencies the amplitude of displacement will still be limited mainly by the stiffness of the neuromast but as the frequency increases a point will be reached where the neuromast resonates. Experiments on the canal neuromasts on the head of the killifish (*Fundulus*) indicate that their resonant frequencies are between about 90 and 180 cycles/sec.[47] Around the resonant frequency the amplitude of forced vibrations is limited mainly by viscosity. Canal neuromasts may be less sensitive than superficial ones near the resonant frequency because the viscous resistance will be higher owing to the slenderness of the canal. They may also tend to have lower resonant frequencies because of the inertia of the fluid in the canal that must move with the cupula. At frequencies above the resonant frequency, amplitude is limited by inertia, but there is no evidence that lateral line neuromasts can respond to frequencies much above 200 cycles/sec.

The main function of the lateral line system is probably the perception of disturbances in the water due to moving objects. Blinded fish can be trained to locate moving or vibrating glass rods so long as the lateral line system is left intact.[36] Electrophysiological experiments indicate that the lateral line neuromasts can detect acoustic near fields at frequencies up to 200 cycles/sec. Many fish which take active food from the bottom may use their lateral line system to locate it. A wriggling worm on a muddy bottom might be impossible to see if its movements stirred up the mud, but the lateral line system could locate it. The lateral line system may also be useful in giving warning of the approach of predators.

Elimination of the lateral line system does not seem to affect the ability of a fish to swim.[36] The ability to keep station against a current does not depend on the lateral line system but on vision. Blinded fish put into flowing water are swept away by the current until they happen to strike the bottom, and only then start swimming against the current. They obtain information about the current from their touch receptors, not from the lateral line system. One would not expect the lateral line system to be much use in keeping station because information about the velocity of water relative to the fish could be no help in keeping station unless the velocity of the water relative to the ground was known.

The lateral line system may help fish to find their way among obstacles in darkness. A number of experiments with blinded or naturally blind fish have shown that they can avoid obstacles without touching them.[36] As a fish swimming at a constant speed approaches an obstacle, the pressure at its snout will rise and the velocity of flow of water over its head will increase because the obstacle obstructs the flow of water displaced by the fish. It is presumably by detecting these changes that the fish avoid the obstacles.

The sensitivity of a sense organ to a stimulus is reduced if the stimulus must be perceived against a background of other stimuli which affect the same neurones. It is difficult to follow a conversation in a noisy room. It is desirable that a sense organ be made selectively sensitive to the particular stimuli which are important to the animal. The lateral line organs seem to be used for perceiving local disturbances in the water rather than the flow of water due to currents or swimming. The canal neuromasts are selectively sensitive to these disturbances. Canal neuromasts would seem for this reason to be more useful than superficial ones. However, superficial neuromasts may be more sensitive than canal ones to frequencies around their resonant frequency.

Lampreys and hagfishes have superficial neuromasts and no canals. Selachians and bony fish usually have both superficial neuromasts and canals, but there are considerable differences among bony fish. Those that swim a good deal or which rest on the bottom in streams usually have an extensive system of canals.[36] They are the species to which the selective sensitivity of the canal organs is particularly important. Bony fish which live in still water and do not swim much have less extensive canals or no canals at all. Among the cyprinoids, for instance, active swimmers such as the roach (*Rutilus*) have extensive canals and so does the stone loach (*Nemacheilus*) which spends a lot of time resting on the bottom in streams, but another loach (*Misgurnus*) which lives in still water has no canals at all. The Australian lungfish (*Neoceratodus*) lives in rivers and has canals but the South American one (*Lepidosiren*) lives in swamps and does not. The angler fish (*Lophius*) has no canals. The herring is an exception to the general rule. It is an active swimmer but has no lateral line canals on its trunk.

Some blind fish which live in perpetual darkness in caves have very large numbers of superficial neuromasts. The gulper eel (*Eurypharynx*), which lives deep in the sea where the light is dim, has groups of neuromasts on the ends of stalks projecting from its body.[5] The macrourids, which live near the bottom in deep water and which, unlike the gulper eel, are probably fairly active, have unusually wide lateral line canals on their heads.[72] The width of the canals probably makes the system unusually sensitive around the resonant frequency by reducing viscous damping. The witch (*Glyptocephalus*) is a flatfish which lives on muddy bottoms, whereas most flatfish live on sand. It has very wide lateral line canals on the lower side of its head.[36] They may help it to find the invertebrates it eats among disturbed mud. The openings of the canals are covered by membranes of skin. These keep the mud out of the canals, but are flexible enough to allow the movements of fluid along the canals which displace the neuromasts.

When a stone is dropped into water a series of expanding concentric ripples is formed. Similar ripples result from gentler disturbances of the surface, for instance by an insect landing on it. Passage of the ripples involves movements of the water up and down, and towards and away from the centre of disturbance. Recent experiments on a species of *Fundulus* and on another cyprinodont suggest that they may use ripples to locate food.[84] Both spend much of their time, at least in aquaria, close under the surface with the tops of their heads only just submerged. The tops of their heads are broad and flat, and, as in other fish, bear neuromasts. In *Fundulus* these are in short lengths of canal but in the other species they are superficial. They are set at various angles and so are sensitive to water movements in any direction. Blinded fish of both species can be trained to swim to the source of a system of ripples. If some of the neuromasts on the head are destroyed the fish fail to respond to ripples from certain directions, or swim in the wrong direction. A fish looking upwards can of course see out of the water, and if it looks at the critical angle it will see the horizon, but vision at angles near the critical angle is marred by chromatic aberration and by partial reflection.

It may well often be easier to detect the ripples from an insect on the surface than to see it.

Inertia and the ear

The lateral line canal neuromasts can be expected to be affected by acceleration and deceleration of the fish as well as by the sorts of stimuli we have been considering.[92] Suppose a swimming fish stops suddenly. The fluid in the longitudinal canals will tend to continue travelling forward, owing to its inertia, and the neuromasts will be deflected anteriorly. Suppose a fish turns. One side of its body will accelerate and the other decelerate, and the neuromasts in the longitudinal canals will be displaced in opposite directions on the two sides of the body. The lateral line system is thus in principle capable of detecting both linear acceleration and rotation of the body. The ears of fish appear to be parts of the lateral line system which have become specialised for these functions. They have sunk into the body and lost their connection with the outside, except through the slender ductus endolymphaticus which remains in some fish. They have been modified in the process. If they had not, they would not work in isolation from the external water. If inertia is to carry water along a canal when the fish accelerates or decelerates, the canal must open to the exterior or at least have a flexible wall, in either of which cases it will be affected by external water movements as well as by movements of the body.

The semicircular canals are specialised receptors for rotation. We have already seen how they work. The fluid is made to flow in them by angular accelerations, but is not affected by linear accelerations or movement of the water outside the body. All vertebrates except the Agnatha have three canals, mutually at right angles, in each ear. Among the Agnatha, the lampreys have two canals in each ear and the hagfishes have only one with two ampullae.[92] All rotations can be resolved into components about three axes, mutually at right angles. A fish could obtain full information about its rotations from three semicircular canals suitably arranged. Six canals (two in each ear) are twice as many as are really necessary. Even lampreys could get full information about rotations from one ear, for though each ear has only two canals their structure is more complex than in other vertebrates and each ear can sense three components of rotation.

As well as semicircular canals, ears contain otolith organs. These are neuromasts with a deposit of calcium carbonate crystals in their cupulae. The calcified cupulae, or otoliths, have a specific gravity of about 2. This enables them to function as receptors for linear accelerations, although they are in an enclosed chamber. When a fish accelerates, the dense otoliths tend to lag behind. Although the lagging otolith drives the fluid surrounding it forward the centre of gravity of the ear as a whole lags because the otolith is denser than the fluid.

Since the otoliths are denser than the fluid surrounding them, they tend to be drawn downwards by gravity. The frequency of action potentials in their neurones is influenced by the posture of the fish as well as by its acceleration. Precisely the same thing happens in our ears; we cannot distinguish between accelerating forward and leaning backward on the basis of information provided by our otolith organs alone. This ambiguity does not of course make otoliths useless. Experiments in which otoliths have been removed show that the utricular otoliths of fish are very important in maintenance of posture.[1]

There are three otolith organs in each ear. Action potentials have been recorded from their neurones in the ray (*Raia*) and they have been shown to be sensitive both to acceleration and to tilting. Two of them, in the sacculus and utriculus, can discriminate between pitching and rolling movements.[1] This has not been yet explained in terms of the orientation of the kinocilia and their orientation does not seem sufficiently varied to account for it. Half are on the dorsal and half on the ventral sides of the sensory hairs.

The otoliths in the sacculus and utriculus of the ray have also been shown to be sensitive to vibrations, which of course involve acceleration first in one direction and then in the other.[1] They are presumably useful in hearing, for a fish exposed to sound will vibrate with the water surrounding it. A great many observations and experiments have shown that fish can hear, but the relative importance of the otoliths, the lateral line system and the touch receptors in the skin is not usually clear.[1,92] Fish ears have no cochlea.

Gas and hearing

Sound involves pressure changes and gas is much more compressible than water. When a bubble of gas is placed in water through which sound is passing it expands and is compressed more than the water. Water in contact with its surface vibrates with greater amplitude than it would if the bubble was not there. The sound vibrations are amplified.

The degree of amplification depends on the frequency of the sound.[11] It may be several thousand times in an acoustic far field at the resonant frequency of the bubble, which is about $300/r$ cycles/sec for a bubble of radius r cm in shallow water. It falls off on either side of the resonant frequency, but is still about 10 times at $60/r$ and $2,000/r$ cycles/sec. We have seen that the ratio of pressure amplitude to displacement amplitude is reduced close to a sound source, where the near field is important. Because of this, a bubble does not amplify sound vibrations so much when it is close to the source of sound.

The amplification is of course not limited to the bubble, for the water surrounding it must move as it pulsates. The bubble sets up an acoustic near field around itself. Any bubble of gas in the body will increase the amplitude of vibration in the ear, but the nearer it is to the ear, the more effect the bubble will have.

The labyrinth fishes (*Anabas*, etc.) take bubbles of air into the dorsal parts of their opercular cavities, where they have air-breathing organs (Chapter 4). These bubbles lie immediately lateral to the part of the cranium which contains the ears. If the cranium had an ordinary stiff bony wall, the pulsation of the bubbles would affect the fluid of the ears far less than the water on their other sides. However, flexible membranous windows have evolved in the cranium, next to the bubbles, so that the ears are affected as much as possible by the acoustic near fields of the bubbles. Experiments in which labyrinth fishes have been trained to respond to sounds have shown that they are much less sensitive to far field sound if the air is removed from their opercular cavities.[11]

Since the swimbladder is filled with gas, it amplifies sound. It is not as good an amplifier as a free gas bubble because it is entirely surrounded by tissues whose viscosity damps its vibrations. The amount of damping can be estimated from records of the sounds which certain fish produce,[11] and also by other more direct methods. Swimbladders probably only amplify the vibrations in a far field at their resonant frequency by some factor in the region of 100, instead of several thousand. At frequencies well away from the resonant frequency, viscous damping is much less important and a swimbladder is probably nearly as good as a bubble.

A swimbladder will improve the hearing of any fish to some extent. However, typical swimbladders do not reach particularly near the ear. Many groups of fish have evolved anterior diverticula of the swimbladder which make contact with the ear through openings in the skull. They include the herring and its relatives (Clupeiformes), in which the diverticula consist of very fine ducts ending in vesicles inside the cranium.[2] It seems likely that the acoustic near field of the main body of the herring swimbladder has relatively little effect on the ear, but that sensitivity to high frequencies is greatly increased by the near field of the vesicles (since the vesicles are small, their resonant frequency is high). The Mormyriformes have rather similar vesicles, but their connection with the main body of the swimbladder is lost in adults.[2]

The Ostariophysi, which include the majority of freshwater fish, have a quite different connection between the swimbladder and the ear.[11] It is known as the Weberian apparatus and is their distinguishing feature. It may be largely responsible for their success, but the fright reaction (Chapter 5) is also peculiar to them and is shown by most of them.

The basic principle of the Weberian apparatus is shown in fig. 16E. The swimbladder or an anterior part of it has a double wall, which is shown in section. The inner wall (tunica interna) is complete but the outer one (tunica externa) has a longitudinal slit dorsally. They are attached to each other only by delicate oily connective tissue. When the swimbladder is exposed to sound, it expands and contracts. When it expands the tunica interna must stretch but the tunica externa need not; it can slide over the surface of the tunica interna, widening the slit as shown in the figure. Movements of the edge of the slit are transmitted to the ear by a chain of small bones. Elastic connections between these ossicles and the vertebral column bring the edges of the slit together again when the swimbladder contracts.

Figure 16F is a diagrammatic dorsal view of the swimbladder and Weberian apparatus of a carp or characin (Cypriniformes). The apparatus of catfish (Siluriformes) is essentially similar but their swimbladder lacks the constriction. To avoid making the swimbladder excessively large or the Weberian apparatus excessively small, the figure has not been drawn to a uniform scale.

The anterior parts (right) are drawn to a much larger scale than the posterior ones. To show the way in which the saccular otoliths move it has been necessary to draw them as they would appear in transverse section.

There are three ossicles on each side attached to each other by ligaments. They have flexible attachments to the vertebrae. The most posterior ossicle, for instance, is attached to the third vertebra by a strip of bone or cartilage which acts as a leaf spring, returning it to its original position when it is displaced and released. There is also an elastic ligament between this ossicle and the fourth vertebra.

The posterior ossicles are attached to the tunica externa of the swimbladder at the edges of the slit. When they move, the other ossicles move with them because of the ligaments. The anterior ossicles are incorporated in the walls of a median fluid-filled cavity which enters the cranium through the foramen magnum. This cavity encloses a thin-walled diverticulum of a canal which connects the sacculi. When the swimbladder expands, the edges of the slit in the tunica externa move apart, the ossicles swing anteriorly, fluid is forced into the sacculi and the saccular otoliths are displaced (fig. 16F). When the swimbladder contracts, the reverse happens because the posterior ossicles are drawn together by the elasticity of their attachments to the vertebral column. The movements of fluid in and out of the sacculi are possible because the sacculi have flexible regions in their walls.

The various otoliths would vibrate when the fish was exposed to sound, even if there were no Weberian apparatus. The Weberian apparatus increases the amplitude of vibration of the saccular otoliths, and so makes the fish more sensitive to sound. The sensitivity of fish to sound can be investigated by training them to respond to sound. They can, for instance, be trained to come for food when a loudspeaker emits a sound. The pitch and intensity of the sound can be varied to establish the limits of what the fish can hear. It is, of course, necessary to take many precautions in investigations of this sort. In particular, one must be sure that the fish really are responding to the sound, and not merely to the spectacle of an expectant scientist standing ready to record their behaviour. Many investigations of this sort have been made but many of the results are hard to interpret because the stimuli have been an unspecified mixture of acoustic near and far fields. However, it seems clear that the Ostariophysi hear exceptionally

E

well, and that their sensitivity to sound is much reduced when the swimbladder is deflated or when the posterior Weberian ossicles are removed.

Comparisons between the hearing abilities of fish and people are awkward because of the difference between sound in air and sound in water. The ratio of pressure amplitude to displacement amplitude is much greater in water than in air. If sounds in air and water involving equal energy flux are considered to be equally loud, Ostariophysi seem able to hear far field sound about as well as we can. However, they seem unable to sense its direction. One would not expect a pressure receptor like the Weberian apparatus to give an indication of the direction of sound.

When a fish changes its depth, the volume of its swimbladder changes (Chapter 3). A tiny change of depth would change the volume of the swimbladder as much as an extremely loud sound. Quite a small change might make some of the ligaments of the Weberian apparatus slack and put the apparatus out of action, were it not for the properties of the tunica externa. It consists mainly of needle-like crystals of a peculiar form of collagen. These are held together only by a delicate network of elastin. Since the elastin is extensible, the tunica externa is extensible, but the needles of collagen make it very viscous. When a force is applied to it, it stretches very slowly. It probably yields very little when the tunica interna expands and contracts under the influence of sound, but when the fish swims to a new depth it probably adjusts its dimensions slowly to the new size of the swimbladder so that the width of the slit returns to normal. The tunica externa of a living ostariophysan fish is very different from the tough structure found in preserved ones, because preservation converts the collagen crystals to fibres.

The Weberian apparatus is intricate and there is nothing much like it in fish outside the Ostariophysi. It is not at all clear how it evolved. The structure of the ossicles, both in adults and in embryos, seems to show that the two anterior ossicles on each side have evolved at least in part from parts of the first two vertebrae. The posterior ossicles seem to be parts of the third vertebra with rudimentary ribs attached to them. Apart from these rudiments, the first four vertebrae have no ribs. In many teleosts outside the Ostariophysi the first few vertebrae have reduced ribs or no ribs, and in some the swimbladder adheres to the vertebrae or the bases of the ribs above it. These observations can be taken as hints about the sort of structure from which the Weberian apparatus may have evolved, but give no indication of the origin of the connection with the ear.

The swimbladder lies in the body cavity and in most Cypriniformes, as in other teleosts, it is separated from the water surrounding the fish by a thick wall body. In the catfish (Siluriformes) and in some Cypriniformes there is a gap in the muscles of the body wall on each side. The swimbladder extends into the gap and is there separated from the water surrounding the fish only by the skin. This presumably reduces the viscous damping of the swimbladder's pulsations and makes the Weberian apparatus more sensitive to sounds near the resonant frequency of the swimbladder.

Many teleosts have reduced swimbladders or no swimbladders at all (Chapter 3). Many Ostariophysi such as the stone loach (*Nomacheilus*) and the sucking catfish (*Plecostomus*) have reduced swimbladders, but none seem to have lost their swimbladders altogether. A reduction in the size of the swimbladder raises its resonant frequency and so changes the range of frequencies which it amplifies, but does not necessarily reduce the amplification. It is probably the value of the Weberian apparatus that has prevented the loss of the swimbladder.

Electric receptors

The sense organs which have been described so far are all sensitive to mechanical stimuli of one sort or another, and are believed to serve as mechanoreceptors. Selachians and various teleosts have sense organs which appear to be related to them, but which seem to serve primarily as electric receptors.

The most conspicuous of these electric receptors are the ampullae of Lorenzini of selachians.[36] Each ampulla is filled with jelly and opens to the surface through a duct, also filled with jelly. The ducts can easily be seen by stripping the skin from a selachian head. Some of them are several or even many centimetres long so that, though the ampullae are grouped in a few small clusters, the openings of the ducts are very widely distributed. Dogfish and sharks have openings scattered all over the head, and rays have some on the pectoral fins as well.

Catfish, Gymnotoidei and Mormyriformes have organs rather like the ampullae of Lorenzini, but they have much shorter ducts (except in the catfish *Plotosus*), and they are often scattered all over the body. Electron micrographs of the ampullary organs of a catfish show that the sensory cells have only short microvilli extending into the lumen of the ampulla.[75] There are no long microvilli like the stereocilia of neuromasts, and no kinocilia. Sensory neurones synapse with the sensory cells, just as neurones synapse with the hair cells of neuromasts. Gymnotoidei and Mormyriformes also have tuberous organs which seem to have no external opening but resemble the ampullary organs in other respects.[90]

The neurones of the ampullae of Lorenzini, like those of neuromasts, carry a constant succession of action potentials when no stimulus is applied. The frequency of the action potentials can be altered by a remarkable variety of stimuli. It increases when the ampullae are cooled and decreases when they are warmed. Distinct responses to changes of as little as 0·1°C have been recorded. Ordinary lateral line neuromasts respond similarly, but are far less sensitive. The ampullae seem capable of serving as temperature receptors, but there is no apparent reason why a temperature receptor should have a long duct filled with jelly or why it should be buried deep in the body, as some of the ampullae are. The frequency of action potentials can also be altered by a very gentle touch with a hair or a fine nylon filament on the opening of the duct. It is usually increased by a touch but in some ampullae it is decreased. Other mechanical stimuli such as a jet of water or movement of the skin are also effective. The ampullae are also sensitive to electrical stimuli. The frequency of action potentials is increased when the opening of the duct is made negative to the ampulla and decreased when it is made positive. A gradient of 10^{-6} volt/cm is enough to alter the frequency by 10%. All neurones are sensitive to electrical stimulation (a property which has been extremely useful to neurophysiologists) but the ampullae are quite remarkably sensitive. They are also sensitive to small changes in the salinity of the water in contact with the openings of their ducts. This seems to be a direct consequence of their electrical sensitivity.

If the ampullae of Lorenzini serve as electrical receptors, their ducts and jelly can be understood. The electrical conductivity of the jelly is almost as high as that of sea water and twice as high as that of the body fluids of the fish. If there is a potential difference between the ends of a duct, more current per unit cross sectional area will flow along it than through the surrounding tissue. The jelly will concentrate the current.

Electrophysiological experiments seem to show that the ampullae are capable of serving a variety of functions. Other experiments seem to show that they are actually used as electric receptors. It has been shown by recording their heart beats that selachians perceive very weak electric stimuli, including gradients of as little as 10^{-8} volts/cm. When the stimulus is applied the frequency of the heart beat changes. They no longer respond when the nerves to the ampullae are cut. It has even been shown in this way that *Raia* can detect the action potentials of the respiratory muscles of a resting plaice (*Pleuronectes*) at a range of 5–10 cm.[62] To make certain that non-electric stimuli were not involved, the plaice and the ray were put in separate aquaria. The action potentials of the plaice were recorded by electrodes in its aquarium, passed through an amplifier so that their size could be adjusted to simulate various distances from the ray, and delivered by electrodes into the ray's aquarium. The ampullae of Lorenzini may well help rays and other bottom-feeding selachians to find the animals on which they feed (which do not normally include plaice).

No teleost has been shown to be quite as sensitive to potential gradients as rays are, but *Gymnarchus* (Mormyriformes) has been trained to swim to a feeding trough in response to gradients of about 10^{-7} volts/cm.[69]

Among the teleosts known to have ampullary organs only the catfish *Plotosus* has ones with long ducts and only it is marine. If a jelly-filled duct is to concentrate current flow, the jelly must have a higher electrical conductivity than the body fluids of the fish. This can be achieved by including a higher concentration of ions in it than in the body fluids. This presents no problem for a marine fish, since the jelly can be given the same ionic concentration as sea water, but it is hard to see how a freshwater fish could maintain a high concentration of ions in an open jelly-filled duct. The ducts of the electric receptors of freshwater fish are either short or have been lost.

Electric organs

We have seen that electric receptors can detect muscle action potentials. Many fish that possess them also have modified muscles which have lost their power of contraction and serve only to generate electrical potentials.[32] In most cases these are modified longitudinal swimming muscles but in the electric ray (*Torpedo*) they are branchial muscles. The Mormyriformes, the Gymnotoidei and the electric catfish (*Malapterurus*) are freshwater teleosts and have electric organs which produce action potentials. The ray (*Raia*), the electric ray and one genus of marine teleosts have electric organs which do not produce action potentials; their discharges are end-plate potentials. The sharks and dogfish, and catfish other than *Malapterurus*, have electric receptors but no electric organs.

The cells of electric organs are not slender like muscle fibres, but are flat plates. If action potentials occurred simultaneously in the two faces of a plate they would cancel each other out and the fluids on either side of the plate would remain at the same potential. If action potentials occur at slightly different times on the two faces potential differences appear across the plate, first in one direction and then in the other. This is what happens in some electric fishes such as *Gymnotus*, in which action potentials start on one side of each plate and spread to the other. In other fishes such as *Malapterurus* action potentials start at about the same time in both faces of each plate, but last longer in one face than in the other. In yet others, such as *Electrophorus*, only one face of each plate is subject to action potentials.

Action potentials are small; they vary between about 0·1 and 0·15 volts. End-plate potentials are even smaller. In electric organs large numbers of plates are arranged face to face so that potential differences across them are added together, as with batteries connected in series. The nervous system is so organised that action potentials or end-plate potentials can be produced simultaneously in all the plates of an electric organ. Many electric fishes have weak electric organs with 60–200 plates in each series, but large *Electrophorus* and *Malapterurus* have thousands of plates in series and produce potentials of several hundred volts. The electric organs of *Electrophorus* make up some 40% of the volume of the fish.

The functions of electric organs vary from fish to fish. Organs that produce high voltages are used for defence and in attacking prey, but these functions depend on the voltage being high, and strong electric organs must have evolved from weak ones with other functions. Various Mormyridae seem to use the potentials produced by their weak electric organs as threat signals, much as birds use song. Action potentials of ordinary muscles may well have become important as threat signals before electric organs evolved if the Mormyridae evolved electric receptors first.

Mormyriformes and Gymnotoidei also use their electric organs to locate objects around them. They produce electrical pulses continuously, at frequencies which vary greatly from species to species. Among the Gymnotoidei, *Hypopomus* and *Electrophorus* produce only 1–5 pulses/sec when they are resting, and up to about 20/sec when excited, but *Sternarchus* produces 650–1,000 pulses/sec. *Electrophorus* does not use its strong electric organs to produce these regular pulses, but other organs which produce much smaller potentials. *Gymnarchus* produces about 300 pulses/sec.[46]

The electrical pulses set currents flowing in the water round the fish. The pattern of current flow is modified by objects whose electrical conductivity is different from that of the water. An object whose conductivity is higher than that of the water concentrates the current, and so increases the current flowing through nearby parts of the body of the fish. An object whose conductivity is lower than that of the water has the reverse effect. Objects in the water can therefore alter the currents which flow during each pulse through individual electric receptors in the fish. Action potentials have been recorded from neurones serving the electric receptors of various gymnotoids, and it has been shown that their frequency can be altered by bringing pieces of metal or paraffin wax near the fish.[46] *Gymnarchus* has been shown to be able to detect a glass rod of 2 mm diameter hidden in a porous pot.[69] It was trained to take food presented beside a pot filled with water, and to reject food presented beside a pot containing water and the glass rod. Experiments with pots with other contents confirmed that the pots were being distinguished by their electrical conductivities.

The Mormyriformes and Gymnotoidei live in fresh water. Animals in the water around them therefore have a higher conductivity than the water. Rocks have a lower conductivity than the water. Their electric organs can probably help the fish both to locate predators and food, and to find their way about. The Mormyriformes live in very turbid water and the Gymnotoidei are only active at night, so vision can be relatively little use for these purposes.[69]

Gymnarchus swims with its body held straight by undulating its very long dorsal fin (Chapter 2). Gymnotoidei do the same thing with a long anal fin. This peculiar manner of swimming may make the electric location of objects easier. If the fish swam in the usual way by moving their tails from side to side, the electric organs would move from side to side and their positions relative to the electric receptors anterior to them would keep changing. This would complicate analysis by the brain of information from the receptors. Swimming with the body straight may enable a brain of limited complexity to make a more complete analysis.[68]

7

INTERACTIONS

The purpose of this short chapter is to make some general points which are implicit in the preceding chapters. They are concerned with interaction between different functions or consequences of a single structure.

It is often necessary to consider several implications of a single aspect of fish design. If one is considering the densities of fish (Chapter 3) it is not enough to relate them to the energy needed for swimming; if one did no more than this one would conclude that neutral buoyancy should always give a selective advantage. Many teleosts have reduced swimbladders or have lost the swimbladder entirely and have negative buoyancy. This seems usually to be because negative buoyancy gives the advantage of frictional forces which tend to hold the fish in place when it rests on the bottom. For many fishes the balance of advantage favours neutral buoyancy, but for others with different habits it favours negative buoyancy. If one is discussing the gill areas of different species of fish (Chapter 4), it is not enough to consider their respiratory needs. One must also consider the osmotic work made necessary by diffusion of water and ions through the gill epithelia. A higher gill area allows more rapid uptake of oxygen, but makes more osmotic work necessary. The ideal gill area for a particular fish is a compromise value, and depends on the habits of the fish. Other compromises, concerning the size of the eggs and of the ovary, were discussed in Chapter 1.

The evolution of a structure often results in modification of other structures which serve very different functions. The organs of fish are generally neatly packed in the body, and it is desirable that they should be. Neat packing results in a relatively small body which is subject to less drag in swimming than a larger body containing the same organs less neatly packed. The liver is a convenient packing organ whose shape seems to have little or no influence on its functioning. It is moulded to the shapes of the gut ventral to it and (in bony fish) the swimbladder dorsal to it, whatever those shapes may be.

The parts of the head are usually just as neatly packed as the viscera. A striking exception is provided by a blind cave fish, *Anoptichthys*, in which the eye is reduced to a vestige but space for an almost full-sized eye remains, filled only by connective tissue which has no apparent function. In most fish the eye is in close contact with the cranium dorsal and medial to it, and with the jaw muscles (adductores mandibulae) ventral and posterior to it. The tendons of the jaw muscles have to be arranged in a particular way so that the eyeball is not compressed when the muscles contract.[8] Changes in the size of the eye in the course of evolution have made necessary changes in the shape of the cranium, jaw muscles and suspensorium, and even of the brain.[10] The piranha has exceptionally large jaw muscles which make its vicious bite possible (Chapter 5). In typical teleosts the jaw muscles lie lateral to the suspensoria and the gills lie median to them (fig. 9A). The unusually big head of the piranha has more space than usual both lateral to the suspensoria and medial to them. Not only do the muscles fill the large space lateral to the suspensoria, but parts of them have extended into the space medial to the suspensoria. The dorsal ends of the gills have consequently been shifted posteriorly, and the cranium has evolved a curious ventral keel to provide origins for some of the smaller muscles displaced by the jaw muscles.[8]

Structures can acquire new functions in the course of evolution, and may lose their original functions. A variety of organs have been modified in different teleosts to serve as aerial respiratory organs (Chapter 4). Muscles have been modified in several groups of fish to become electric organs (Chapter 6). The maxillae of most teleosts form part of the mechanism of the upper jaw (Chapter 5) but those of catfish (Siluriformes) serve only to support a pair of sensory barbels. They and their barbels are moved by muscles which have evolved from parts of muscles which move the jaws and suspensoria of other teleosts.[10] Many other teleosts have evolved barbels, though none make the same use of the maxilla. Gurnards (Triglidae) and some Anabantoidei have modified fin rays which serve the same purpose.

These examples illustrate the tendency for similar changes of function to occur in different groups of fish. A particularly good example of this is provided by the evolution of spiny fin rays. Acanthopterygians have spiny rays in their dorsal, pelvic and anal fins (Chapter 2). These differ from the ordinary rays from which they have apparently evolved in being stiff and sharp. They may have some function in locomotion, for instance in the control of stability, but they probably also help to protect their possessors against predators. The spines of typical acanthopterygians are not particularly robust, but some acanthopterygians and some fish from other groups have stronger spines whose protective function is more certain. They include the stickleback (*Gastorosteus*) and the catfish.[10] The stickleback may be descended from typical acanthopterygians. It has three large spines at the anterior end of its dorsal fin and one at the anterior end of each pelvic fin. The catfish are certainly not acanthopterygians. They have a large spine at the anterior end of the dorsal fin and of each pectoral fin.

Since most predators that eat fish swallow them whole, a few strong spines on a fish are enough to ensure that predators cannot swallow it without pain or injury. It has been shown that pike and perch learn not to attack sticklebacks after a few experiences of their spines. The bitterling (*Rhodeus*, Cyprinoidei) has a close relative with spines in its dorsal fin. In a region where the spiny bitterling is more common than the bitterling, it is much less often found in the stomachs of predators. Predators apparently eat the bitterling in preference to the spiny bitterling.[1a]

If a spine is to be effective in protecting a fish it must not be easily pressed down against the body by a predator's jaws. Sticklebacks and catfish have not only evolved spines independently; they have also evolved mechanisms for locking the spines in the erect position. The locking depends on friction, just as knots do. The harder one pulls on ropes which are knotted together, the more tightly they are pressed together and the greater the friction between them. Provided they are reasonably rough the knot will never slip, no matter how great the tension. The pectoral and dorsal spines of catfish lock in different ways, which can be related to differences between the pectoral and dorsal fins of more typical teleosts. The spines of sticklebacks are different again. In each case, however, locking depends on rough surfaces which jam together when the tip of the spine is pressed. The pectoral spines of catfish, for instance, have a curved flange which fits into a curved groove in the pectoral girdle. Flange and groove have rough surfaces. Their shape is such that the spine has to be rotated about its long axis as it is erected and depressed. The spine can easily be erected and depressed if it is rotated at the same time, and the muscles used to erect and depress it have this effect. If the tip of the spine is simply pressed, however, the flange is forced out of alignment with the groove and jams in it. The greater the force, the more tightly the rough surfaces of the flange and groove are pressed together and the greater the friction between them. The spine cannot be depressed by pressing on its tip, unless the force is great enough to break either the spine or the girdle.

The effectiveness of the spines depends on the bones to which they are locked being strong and firmly mounted in the body. In sticklebacks firm mounting has been achieved by the evolution of large plates of bone fused to the pelvic girdles and to the radials which articulate with the dorsal spines. These plates of bone lie in the skin and serve also as armour. A large number of peculiarities in the skeletons of catfish can be explained as adaptations for firm mounting of the bones to which their spines are locked. The two halves of the pectoral girdle are sutured together ventrally. They are exceptionally firmly attached to the cranium because of various modifications, including the conversion of a ligament to a strut of bone. The dorsal fin starts close behind the head and its anterior radials are attached firmly to the anterior vertebrae, which in turn are attached immovably to the cranium. In addition, there is often a direct bony connection between the radials and the cranium.

The pelvic spines of sticklebacks and the pectoral spines of catfish have acquired an additional function. The rough surfaces involved in the friction-locking devices can be rubbed together as the spines are erected and depressed, producing sounds. The sounds may possibly serve as warnings to predators, like the rattle of the rattlesnake. Anything that helps predators to learn to recognise it as spiny is an advantage to a spiny fish because it is more likely to be injured if a predator seizes it and spits it out than if the predator leaves it alone.

New designs can only evolve by modification of existing designs. Evolution of structures with new functions depends on the opportunities offered by existing structures. This is why similar changes have so often occurred in different groups of fishes. The parallel evolution of defensive, sound-producing fin spines by sticklebacks and catfish is striking, but it is only one example of a very common phenomenon.

Studying fish design involves trying to understand the structure of fish in terms of selective advantages. The student must approach his material in the manner of historians and critics. He must never allow his mathematics and physics to fool him into thinking that his conclusions are unassailable. He must consider so many interacting structures and functions that he is always in danger of overlooking an important point. This chapter is intended partly as a warning.

APPENDIX: A CLASSIFICATION OF THE FISHES

This classification gives the systematic positions of the fishes mentioned in this book. In most cases it goes down to the level of orders, but suborders of two very large orders are given. Extinct groups below the level of classes are omitted, as are many surviving orders whose members are not mentioned in the book. The descriptive notes should not be taken as definitions.

CLASS AGNATHA

The Agnatha are the most primitive of all vertebrates and are so different from the others that they are often put in a superclass by themselves, with all the other vertebrates grouped in a superclass Gnathostomata. They have no jaws. Their gill skeleton is peculiar, at least in the modern forms; it is a flexible network of cartilage external to the gills instead of a series of jointed bars internal to them. The earliest known fossil vertebrates are Agnatha. Most of the fossils had fairly short bodies and thick bony scales, and many had their heads covered by bony carapaces, but modern Agnatha are eel-shaped and have no bone. There are several extinct orders and two modern ones which are:

Order Petromyzoniformes, which consists of the lampreys, such as .*Petromyzon*. Adults have suckers round their mouths which they use for attaching themselves to stones and, in some species, to other fish whose blood they suck.

Order Myxiniformes, which consists of the hagfishes such as

Order Myxiniformes, which consists of the hagfishes such as *Myxine*. They do not have suckers, but have toothplates which can be moved in and out of the mouth (Chapter 5).

CLASS ACANTHODII

This is an extinct class of fishes which had a strong spine in front of every fin except the caudal fin. They used to be put in the same class as the Arthrodira because it was believed that the two groups had autostylic jaw suspension: i.e. that the palatoquadrate articulated with the cranium and received no support from the hyomandibular. However, it has recently been shown that the hyomandibular was after all involved in the suspension of the jaws of Acanthodii.[74]

CLASS ARTHRODIRA

An extinct class of fishes often called placoderms. Most had the head and pectoral region covered by thick dermal bone.

CLASS HOLOCEPHALI

This class includes *Chimaera*. It has often been joined in a single class with the Selachii, but the similarities between the two groups are superficial, and a fossil has recently been discovered which seems to show that the Holocephali evolved from a group of Arthrodira which were much too specialised to have given rise to the Selachii as well.[78] The most characteristic feature of the Holocephali is the fusion of the palatoquadrate cartilage to the cranium.

CLASS SELACHII

This class includes the sharks and rays. Nearly all of them are marine. Their skeletons are cartilaginous, their scales are of the placoid type and they do not have swimbladders or lungs. The males have a pair of claspers, used as intromittent organs, formed from the pelvic fins. (Holocephali have similar claspers but have two pairs.) There are two extinct subclasses and one surviving one which is:

SUBCLASS EUSELACHII

Superorder Pleurotremata, which consists of sharks and dog-fishes.

Order Heterodontiformes

Order Hexanchiformes, with six or seven pairs of gill slits instead of the usual five.

Order Lamniformes, including the leopard shark (*Triakis,* fig. 2), the dogfish (*Scyliorhinus*), the tope (*Galeorhinus*), the smooth hound (*Mustelus*) and the porbeagle shark (*Lamna*).

Order Squaliformes, including the piked dogfish (*Squalus*).

Superorder Hypotremata

Order Rajiformes   The rays including *Raia* (fig. 2), *Rhinobatos* and *Myliobatis.*

Order Torpediniformes   The electric rays, including *Torpedo.*

CLASS OSTEICHTHYES

The Osteichthyes are often referred to as the bony fish (which is the meaning of their name), although many extinct members of other classes also had bone. Most of them have a lung or a swimbladder.

SUBCLASS DIPNOI

The Dipnoi are the lungfishes. They have characteristic tooth-plates (fig. 14E). The arrangement of the dermal bones in their skulls is very different from the arrangement in other Osteichthyes. There are several extinct orders and two surviving ones:

Order Ceratodiformes, whose only surviving member is the Australian lungfish (*Neoceratodus*).

Order Lepidosireniformes, which includes the African and South American lungfishes (*Protopterus* and *Lepidosiren*, respec-tively).

SUBCLASS CROSSOPTERYGII

The crossopterygians have a movable joint between the anterior and posterior parts of the cranium. There are several extinct orders including the Osteolepiformes, from which the amphibians evolved, and one surviving order:

Order Coelacanthiformes, the coelacanths including *Latimeria* which is the only known living crossopterygian.

SUBCLASS BRACHIOPTERYGII

Polypterus and *Calamoichthys*, which live in fresh water in Africa, are the only known members of this subclass. They differ from other bony fish in the structure of their pectoral and dorsal fins. They have very thick scales, like those of the earliest fossil bony fish. Some zoologists include them in the Palaeoniscoidei.

SUBCLASS ACTINOPTERYGII

The actinopterygians are the teleosts and their immediate allies. They lack the characters which have been described as characteristic of the other subclasses of bony fish. They have been the dominant group of fish in fresh water since the Carboniferous period, and in the sea since the beginning of the Mesozoic.

Infraclass Palaeoniscoidei This extinct group includes the most primitive actinopterygians. They had thick scales and heterocercal tails, they had more rays than segments in their fins and their maxillae were sutured to the dermal bones of the cheek.

Infraclass Chondrostei, including the sturgeon and sterlet (*Acipenser*, fig. 2) and their allies. They are specialised in some respects, such as in having little bone and small ventral mouths. They are primitive and resemble the palaeoniscoids in other respects; they have heterocercal tails and they have more rays than segments in their fins. The palaeoniscoids are often included in the Chondrostei but are given a separate infraclass in this classification.

Infraclass Holostei The holosteans evolved from the palaeoniscoids and give rise to the teleosts. They have more symmetrical tails than palaeoniscoids and their maxillae are movable (Chapter 5). Only two modern genera are known and they are put in separate orders. Both live in fresh water in America.

Order Amiiformes, including the bowfin (*Amia*).

Order Lepisosteiformes, including the garpike (*Lepisosteus*, fig. 2).

Infraclass Teleostei The classification of the teleosts which follows was introduced and explained by Greenwood, Rosen, Weitzman and Myers.[45]

Superorder Elopomorpha Teleosts which have a leptocephalus larva.

Order Elopiformes The tarpons, etc.

Order Anguilliformes The eels, including *Anguilla* and *Eurypharynx*.

Order Notacanthiformes

Superorder Clupeomorpha Diverticula of the swimbladder extend into the cranium and end in a characteristic arrangement of bullae round the ears. Nearly all the clupeomorphs are marine.

Order Clupeiformes The herring (*Clupea*), the anchovy (*Engraulis*), etc.

Superorder Osteoglossomorpha The parasphenoid and tongue bear the biggest teeth in the mouth. There are characteristic rods at the base of the second gill arch. All the osteoglossomorphs live in fresh water.

Order Osteoglossiformes, including *Notopterus*.

Order Mormyriformes *Gymnarchus* and the Mormyridae. They are African fish with electric organs.

Superorder Protacanthopterygii This superorder probably gave rise to all the superorders of teleosts which remain to be described. Most of its members have an adipose fin which has no fin rays and can be seen behind the main dorsal fin of *Salmo* in fig. 1. A similar fin is present in many Ostariophysi. The Protacanthopterygii includes several orders of which the most important is:

Order Salmoniformes, including the salmon and trout (*Salmo*, fig. 1), the Pacific salmon (*Onchorhynchus*), *Salvelinus*, the pike (*Esox*, fig. 1), *Synodus* and the lantern fishes (Myctophidae).

Superorder Ostariophysi Weberian ossicles (Chapter 6) connect the swimbladder to the ear. Most species show the fright reaction (Chapter 5). The superorder includes the majority of the freshwater fishes of the world but has very few marine members.

Order Cypriniformes

Suborder *Characinoidei* These are African and South American fish whose teeth are often specialised for cutting. They include the piranhas (*Serrasalmus*), *Myleus*, *Erythrinus* and *Anoptichthys*.

Suborder *Cyprinoidei*, with protrusible toothless jaws and characteristic pharyngeal teeth (fig. 14A–D). This is a very large and widely distributed suborder. It includes the carp

(*Cyprinus*, fig. 1), goldfish (*Carassius*), gudgeon (*Gobio*), bream (*Abramis*), roach (*Rutilus*), dace (*Leuciscus*), orfe (*Idus*), rudd (*Scardinius*), tench (*Tinca*), bitterling (*Rhodeus*), Chinese grass carp (*Ctenopharyngodon*), *Labeo*, *Gyrinocheilus*, *Gastromyzon* and the loaches (*Nemacheilus*, *Misgurnus*, etc.)

Suborder Gymnotoidei   These are the South American electric fishes such as *Gymnotus*, *Sternarchus*, *Hypopomus* and the electric eel (*Electrophorus*).

Order Siluriformes, the catfish. Most of these are nocturnal fish with sensory barbels (including a pair supported by modified maxillae) and with spines at the anterior ends of their dorsal and pectoral fins. They include *Clarias*, *Hoplosternum*, *Malapterurus*, *Plecostomus* and *Plotosus*.

Superorder Paracanthopterygii   These are marine fish which often have one or more of the characteristics of the Acanthopterygii but seem to have evolved from the Protacanthopterygii independently. Their jaws are not protrusible or only slightly protrusible and the arrangement of their jaw muscles is characteristic.

Order Percopsiformes

Order Batrachoidiformes, including the toadfish (*Opsanus*).

Order Gobiesociformes

Order Lophiiformes, the angler fishes, including *Lophius*.

Order Gadiformes, including cod, etc. (*Gadus*, fig. 1), hake (*Merluccius*) and the Macrouridae.

Superorder Atherinomorpha   Many members of this superorder show sexual dimorphism. Many have protrusible jaws but the mechanisms of protrusion are different from those of Cyprinoidei and Acanthopterygii. There is only one order.

Order Atheriniformes, which includes the flying fishes and the toothcarps (Cyprinodontoidei, including the killifish, *Fundulus*).

Superorder Acanthopterygii   This is the largest superorder of fishes. There are usually spiny rays in the dorsal, pelvic and anal fins. The pelvic fins are close to the pectoral girdle (see *Perca* and *Tilapia* in fig. 1). The jaws are usually protrusible and the mechanism of protrusion is characteristic of the superorder (Chapter 5). The orders include:

Order Beryciformes

Order Zeiformes, including the John Dory (*Zeus*).

Order Gasterosteiformes, including the stickleback (*Gasterosteus*) and the sea horse (*Hippocampus*, fig. 1).

Order Scorpaeniformes, with a characteristic arrangement of the superficial bones of the cheek. It includes the miller's thumb (*Cottus*), the lumpsucker (*Cyclopterus*) and the gurnards (Triglidae).

Order Perciformes This is a very large order. Some of the suborders are given below.

Suborder Percoidei, including the groupers (*Epinephalus*), Nile perch (*Lates*), perch (*Perca*, fig. 1), *Lepomis*, the Cichlidae (which include the angelfish (*Pterophyllum*) *Tilapia* (fig. 1), *Haplochromis, Lethrinops* and *Astatoreochromis*) and the butterfly fishes (Chaetodontidae).

Suborder Labroidei, the wrasses, including *Epibulus*.

Suborder Blennioidei, the blennies, including *Blennius*.

Suborder Callionymoidei, including the dragonet (*Callionymus*).

Suborder Gobioidei, the gobies and the mud skipper (*Periophthalmus*).

Suborder Scombroidei, the mackerels (*Scomber*) and tunnies (*Thunnus, Euthynnus*, etc.).

Suborder Anabantoidei, the labyrinth fishes such as *Anabas*.

Order Pleuronectiformes, the flatfishes, including the plaice (*Pleuronectes*), the halibut (*Hippoglossus*) and the witch (*Glyptocephalus*).

Order Tetraodontiformes, including the puffer fishes.

REFERENCES

This list is intended as a starting point for readers who want to know more about the topics discussed in this book, and as a statement of the principal sources from which I have obtained information. It is not by any means an exhaustive bibliography. I have generally listed review articles in preference to primary sources.

SOME USEFUL BOOKS ON FISHES

1 Brown, M. E. (Ed.) (1957) *The physiology of fishes.* New York
1a Gerking, S. D. (Ed.) (1967) *The biological basis of freshwater fish production.* Oxford
2 Grassé, P. P. (Ed.) (1958) *Traité de zoologie* 13. *Agnathes et poissons.* Paris
3 Hardy, A. C. (1959) *The open sea 2, Fish and fisheries.* London
3a Hoar, W. S. and Randall, D. J. (1970–?). *Fish Physiology.* New York
4 Lagler, K. F., Bardach, J. E. and Miller, R. R. (1962) *Ichthyology.* Ann Arbor.
5 Marshall, N. B. (1965) *The life of fishes.* London
6 Norman, J. R. and Greenwood, P. H. (1963) *A history of fishes,* 2nd ed. London
7 Romer, A. S. (1966) *Vertebrate paleontology,* 3rd ed. Chicago
7a Varley, M. E. (1967) *British freshwater fishes.* London
7b Wheeler, A. (1969) *The fishes of the British Isles and North West Europe.* London

OTHER REFERENCES

8 Alexander, R. McN. (1964) *J. Linn. Soc.* (Zool.) **45**: 169–90 (South American Characinoidei)
9 Alexander, R. McN. (1965) *J. exp. Biol.* **43**: 131–8 heterocercal tails)

10   Alexander, R. McN. (1966) *J. Zool., Lond.* **148**: 88–152 (functional morphology of catfish)
11   Alexander, R. McN. (1966) *Biol. Rev.* **41**: 141–76 (physics of swimbladders)
12   Alexander, R. McN. (1966) *J. Zool., Lond.* **149**: 288–96 (jaws of cyprinids)
13   Alexander, R. McN. (1967) *J. Zool., Lond.* **151**: 43–64 (jaws of acanthopterygians)
14   Alexander, R. McN. (1967) *J. Zool., Lond.* **151**: 233–55 (jaws of Atheriniformes)
14a Alexander, R. McN. (1968) *Animal Mechanics.* London
14b Alexander, R. McN. (1969) *J. Zool., Lond.* **159**: 1–15 (pressures in feeding)
14c Alexander, R. McN. (1972) *Symp. Soc. exp. Biol.* **26**: 273–94 (energetics of vertical migration)
15   Bainbridge, R. (1960) *J. exp. Biol.* **37**: 129–53 (speed of fish)
16   Bainbridge, R. (1961) *Symp. zool. Soc., Lond.* **5**: 13–32 (speed and power in swimming)
17   Bainbridge, R. (1963) *J. exp. Biol.* **40**: 23 56 (caudal fin in swimming)
18   Ballintijn, C. M. and Hughes, G. M. (1965) *J. exp. Biol.* **43**: 349–62 (trout respiratory muscles)
19   Barham, E. G. (1966) *Science, N.Y.* **151**: 1399–402 (deep scattering layer)
20   Bateman, J. B. and Keys, A. (1932) *J. Physiol.* **75**: 226–40 (gills and osmoregulation)
21   Beamish, F. W. H. (1964) *Can. J. Zool.* **42**: 355–66 (oxygen concentration and metabolic rate)
22   Beamish, F. W. H. (1966) *J. Fish. Res. Bd. Can.* **23**: 109–39 vertical movements)
23   Beamish, F. W. H. and Mookherjii, P. S. (1964) *Can. J. Zool.* **42**: 161–75 (basal metabolic rate)
24   Blaxter, J. H. S. and Holliday, F. G. T. (1963) *Adv. mar. Biol.* **1**: 261–393 (herring)
25   Bone, Q. (1966) *J. mar. biol. Assoc. U.K.* **46**: 321–50 (red and white muscle)
25a Bone, Q. and Roberts, B. L. (1969) *J. mar. biol. Assoc. U.K.* **49**: 913–37 (density of selachians)
26   Breder, C. M. and Edgerton, H. E. (1942) *Ann. N.Y. Acad. Sci.* **43**: 145–72 (sea horse swimming)
27   Brett, J. R. (1965) *J. Fish. Res. Bd. Can.* **22**: 1491–601 (metabolism of swimming *Onchorhynchus*)
28   Brett, J. R. and Sutherland, D. B. (1965) *J. Fish. Res. Bd. Can.* **22**: 405–9 (metabolism of swimming *Lepomis*)
29   Brightwell, L. R. (1953) *Proc. zool. Soc. Lond.* **123**: 61–4 (dogfish feeding)

30 Brodal, A. and Fänge, R. (Eds.) (1963) *The biology of Myxine.* Oslo

31 Carey, F. G. and Teal, J. M. (1966) *Proc. nat. Acad. Sci. U.S.* **56:** 1464–9 (temperature of tunny muscles)

32 Chagas, C. and Paes de Carvalho, A. (eds.) (1961) *Bioelectro-genesis.* Amsterdam

33 Clarke, C. A. and Sheppard, P. M. (1966) *Proc. R. Soc.* (B) **165:** 424–39 (melanic moths)

34 Corbet, P. S. (1961) *Proc. zool. Soc. Lond.* **136:** 1–101 (food of L. Victoria fish)

35 Denton, E. J. (1970) *Phil. Trans. R.Soc.* (B) **258:** 285–313 (silvery layers in fishes)

35a Denton, E. J., Liddicoat, J. D. and Taylor, W. D. (1972) *J. mar. biol. Assoc. U.K.* **52:** 727–46 (permeability of swimbladder to gases)

36 Dijkgraaf, S. (1963) *Biol. Rev.* **38:** 51–106 (lateral line)

37 Downing, K. M. and Merkens, J. C. (1957) *Ann. appl. Biol.* **45:** 261–7 (survival at low oxygen concentrations)

38 Fänge. R. (1966) *Physiol. Rev.* **46:** 299–322 (physiology of swim-bladders)

39 Flock, A. and Duvall, A. J. (1965) *J. Cell. Biol.* **25:** 1–8 (ultra-structure of neuromasts)

40 Fryer, G. (1959), *Proc zool. Soc. Lond.* **132:** 153–280 (natural history of L. Nyasa fishes)

40a Gibson, R. N. (1969) *J. exp. mar. Biol. Ecol.* **3:** 179–90 (pressure under suckers)

41 Gray, J. (1933) *J. exp. Biol.* **10:** 88–104 (fish swimming)

42 Gray, J. (1968) *Animal Locomotion,* London

43 Greenwood, P. H. (1964) *Proc. R. Instn Gt Br.* **40:** 256–69 (African cichlids)

44 Greenwood, P. H. (1965) *Proc. Linn. Soc. Lond.* **176:** 1–10 *(Astatoreochromis)*

45 Greenwood, P. H., Rosen, D. E., Weitzman, S. H. and Myers, G. S. (1966) *Bull. Am. Mus. nat. Hist.* **131:** 339–456 (classification of teleosts)

46 Hagiwara, S. and Morita, H. (1963) *J. Neurophysiol.* **26:** 551–67 (physiology of electric receptors)

47 Harris, G. G. and van Bergeijk, W. A. (1962) *J. Acoust, Soc. Am.* **34:** 1831–41 (physiology of lateral line)

48 Harris, J. E. (1936) *J. exp. Biol.* **13:** 476–94 (stability of swimming selachians)

49 Harris, J. E. (1937) *Pap Tortugas Lab.* **31:** 171–89 (mechanics of fins)

50 Hartley, P. H. T. (1947) *Proc. zool. Soc. Lond.* **117:** 129–206 (natural history of British freshwater fishes)

51 Hersey, J. B., Backus, R. H. and Hellwig, J. (1962) *Deep Sea Res.*
    8: 196–210 (deep scattering layers)
52 Hertel, H. (1966) *Structure, form, movement.* New York.
53 Hiatt, R. W. and Strasburg, D. W. (1960) *Ecol. Monogr.* **30:** 65–127
    (natural history of coral reef fishes)
54 Hickling, C. F. (1966) *J. Zool. Lond.* **148:** 408–19 (pharyngeal
    teeth of *Ctenopharyngodon*)
55 Hughes, G. M. (1966) *J. exp. Biol.* **45:** 177–95 (dimensions of gills)
56 Hughes, G. M. and Ballintijn, C. M. (1965) *J. exp. Biol.* **43:** 363–83
    (respiratory muscles of dogfish)
57 Hughes, G. M. and Morgan, M. (1973) *Biol. Rev.* **48:** 419–75
    (structure and function of gills)
58 Hughes, G. M. and Shelton, G. (1962) *Adv. comp. Physiol. Biochem.*
    1: 275–364 (fish respiration)
59 Johansen, K. (1962) *Comp. Biochem. Physiol.* **7:** 169–74 (cardiac
    output)
60 Johnson, L. (1966) *J. Fish. Res. Bd. Can.* **23:** 1495–505 (growth of
    pike)
61 Jones, F. R. H. (1952) *J. exp. Biol.* **29:** 94–109 (fin movements
    and depth changes)
62 Kalmijn, A. J. (1971) *J. exp. Biol.* **55:** 371–84 (electric sense of
    sharks and rays)
63 Kays, W. M. and London, A. L. (1964) *Compact Heat Exchangers,*
    2nd ed. New York
63a Kevern, N. R. (1966) *Trans. Am. Fish. Soc.* **95:** 363–71 (feeding
    rate of carp)
64 Klausewitz, W. (1965) *Natur. Mus. Frankf.* **95:** 97–108 (*Rhinobatos*
    swimming. In German)
65 Kuhn, W., Ramel, A., Kuhn, J. H. and Marti, E. (1963) *Experientia*
    19: 497–511 (gas secretion in swimbladder)
66 Le Cren, E. D. and Holdgate, M. W. (Eds.) (1962) *The exploitation
    of natural animal populations.* Oxford
67 Lighthill, M. J. (1960) *J. Fluid Mech.* **9:** 305–17 (mathematics of
    swimming)
68 Lissman, H. W. (1961) in *The cell and the organism* 301–17, Eds.
    J. A. Ramsay and V. B. Wigglesworth. Cambridge (*Gymnarchus*
    swimming)
69 Lissmann, H. W. (1963) *Scient. Am.* **208:** 50–9 (electric location)
70 Magnan, A. (1929) *Annl. Sci. nat.* (Zool.) (10) **12:** 5–133 (various
    measurements. In French)
71 Mann, K. H. (1965) *J. Anim. Ecol.* **34:** 253–75 (energy budget of
    fish population)
72 Marshall, N. B. (1965) *Deep Sea Res.* **12:** 299–322 (macrourids)
73 Matthes, H. (1963) *Bijdr. Dierk.* **33:** 1–36 (jaws and pharyngeal
    teeth of Cyprinidae)
74 Miles, R. S. (1964) *Nature, Lond.* **204:** 457–9 (acanthodian jaws)

75 Mullinger, A. (1964) *Proc. R. Soc.* (B) **160**: 345–59 (catfish electric receptors)
76 Myers, G. S. (1960) *Evolution* **14**: 323–33 (fish of L. Lanao)
77 Ohlmer, W. (1964) *Zool. Jb. (Anat)*. **81**: 151–240 (shapes of fishes. In German)
77a Ørvig, T. (1967) In A. E. W. Miles (Ed). *Structural and ·Chemical Organization of Teeth* **1**: 45–110. New York. (evolution of teeth)
77b Osse, J. W. M. (1969) *Neth. J. Zool.* **19**: 289–392 (perch respiratory and feeding movements)
78 Patterson, C. (1965) *Phil. Trans.* (B) **249**: 101–219 (phylogeny of Holocephali)
79 Pfeiffer, W. (1963) *Experientia* **19**: 113–23 (alarm substances)
79a Rao, G. M. M. (1968) *Can. J. Zool.* **46**: 781–6 (metabolism in different salinities)
79b Rayner, M. D. and Keenan, M. J. (1967) *Nature, Lond.* **214**: 392–3. (roles of red and white muscles)
80 Saunders, R. L. (1961) *Can. J. Zool.* **39**: 637–53 (pressures in respiration)
81 Saunders, R. L. (1962) *Can. J. Zool.* **40**: 817–62 (oxygen uptake at gills)
82 Schaeffer, B. and Rosen, D. E. (1961) *Am. Zool.* **1**: 187–204 (actinopterygian jaws)
83 Schumann, D. and Piiper, J. (1966) *Pflüger's Arch. ges. Physiol.* **288**: 15–26 (work of respiration. In German)
84 Schwartz, E. and Hasler, A. D. (1966) *J. Fish. Res. Bd. Can.* **23**: 1331–52 (perception of surface waves)
85 Smit, H. (1965) *Can. J. Zool.* **43**: 623–33 (metabolism of swimming goldfish)
86 Springer, S. (1961) *Am. Zool.* **1**: 183–5 (shark feeding)
87 Steen, J. B. (1963) *Acta. physiol. scand.* **58**: 138–49 (gas resorption from swimbladder)
88 Steen, J. B. (1963) *Acta physiol. scand.* **59**: 221–41 (gas secretion in swimbladder)
89 Steen, J. B. and Kruysse, A. (1964) *Comp. Biochem. Physiol.* **12**: 127–42 (path of blood in gills)
90 Szabo, T. (1965) *J. Morph.* **117**: 229–50 (anatomy of electric receptors)
91 Tietjens, O. G. (1957) *Applied hydro- and aeromechanics.* New York
92 van Bergeijk, W. A. (1960) *Contrib. Sensory Physiol.* 2 (evolution of ear)
93 Walters, V. (1966) *Bull. mar. Sci.* **16**: 209–21 (*Euthynnus* feeding)
93a Webb, P. W. (1971) *J. exp. Biol.* **55**: 489–540 (two papers on efficiency of swimming)
94 Yazdani, G. M. and Alexander, R. McN. (1967) *Nature, Lond.* **213**: 96–7 (flatfish breathing)

INDEX

Page numbers in italics refer to illustrations